PROFESSOR PICKLE AND THE
OMICRON AFFAIR

PROFESSOR PICKLE AND THE OMICRON AFFAIR

AND OTHER FASCINATING MATH ADVENTURES

JERRY FARLOW

THALES HOUSE PRESS

Interior artwork by Jean Watts

Printed in the United States of America

First Printing, 2016

ISBN: 978-1-5407-6422-5

10 9 8 7 6 5 4 3 2 1

Publisher: Thales House Press

Dedication

To those readers who hate math, just skip the equations.

Contents

Preface

I was slightly baffled a while back when a student came rushing in my office, but instead of trying to entice me to cough up the blueprints of an upcoming exam or reciting a well-prepared oration for turning in a late homework, she asked a rather curious question.

"Do you remember an old student of yours from 25 years ago?" she said.

Normally, I don't remember the names of students I had the previous semester, but for some reason I remembered this one student.

"Why yes, I remember her," I admitted.

"*She's my mom!*" she blurted out.

And to make matters worse, the incident didn't take place last week. *It was twenty years ago!* So to make sure that another student doesn't come rushing in my office and ask me if I remember so-and-so, then blurt out, "*She's my grandma!*" I decided to take the safe route, retire, and, as they say, get the hell outta Dodge.

A short while after retirement I began to see the benefits of retirement: snoozing in my favorite rocker, lots of relaxation, lots of contemplation, and even more *boredom.* Observing that I was slowly going stark-raving mad, my wife suggests that after teaching and doing mathematical research for 50 years, I might write a book of some kind. Just to verify that I had been doing *something* in the past half-century, she made the wry suggestion.

Little did she know that for the past fifty years, I *had* been writing a book, I just didn't know it. Often, after a long week slaving over a hot differential equations class, I would crash out on Friday night and spew out a chunk of words about some mundane, albeit mathematical, topic.

Although fifty years will result in a pile of words, getting the average book editor, no doubt an English Lit major from a humanities college, to get exercised over such a collection of stories, categorized as "math miscellanea" is a task not taken lightly. Once the average editor sees a book so classified, a rejection letter is not far behind. However, there are a few enlightened editors who see the value of presenting mathematics in various shades of grey, so here we are.

Professor Pickle and the Omicron Affair consists of short mathematical adventures ranging from satire to serious mathematics to downright silliness.

Feel free to skip over any story that doesn't fit your fancy or contains too much or too little mathematical minutia. Perhaps, however, there are stories which resonate with your curiosity and you will gain something from the experience. Or if you have nothing better to do, feel free to read the entire book, cover to cover. Enjoy.

Jerry Farlow

Professor Emeritus of Mathematics
University of Maine

$$\Phi\Sigma\Xi\Lambda\Theta\Omega$$

Forward

The happiest day of my life was the day when this little *tome* was finally shipped off to the printers! Maybe some grass will finally get mowed around this place, or maybe someone, whose name I will not divulge, will take out the garbage!

If I have to answer one more question about whether the period goes before or after the quotation mark, I'll go stark raving mad. Or the difference between "affect" and "effect," sheesh, don't math professors ever learn any grammar?

Now that he's done with his *magnum opus*, as he calls it, he'll no doubt migrate back to his usual headquarters in front of the TV, watching week-end football, and demanding the chips and dips keep coming.

My only desire is that the dear reader of his *tour de force*, gets as much enjoyment from it as I do, knowing it's 100% done.

Susan Farlow
Author's wife

ΦΣΞΛΘΩ

About the Author

My publisher told me that this was the place in the book where I should include all the pretentious crap about myself that I could muster. He said just don't end a sentence with a preposition (or proposition) or mix up commonly misused words like "elicit" and "illicit," but other than that to illicit anything I could think of.

————¤¤¤ΞΞΞ¤¤¤————

My writing career began on a dark and stormy night when my famed travelwriter wife suggested our funds were trending low, that I might do my part by writing a best-selling [cough] math book.

I said that might be a good idea since my experience in the writing field was established long ago when I spilled a bottle of writing ink on the dress of my first-grade teacher, Miss. Altman, an innocent accident for which she had no quarter. After that came college and my English professor, Mr. Kerrigan, who gave me a D in Eng Comp 101 for my refusal to follow all those fuddy-duddy old rules about composition and English usage.

But things turned around for me after I became a college professor and began accumulating desk drawers of pedagogical tailings. In my attempt to pass on learned words of wisdom to future generations, I spent weekends rummaging through pages of old notes and lesson plans, summarizing their contents in 1,000-word essays. I was ill at ease over the less-than-Harvard-level of scholarship of my writings, and so I published them anonymously in various less-than-Harvard-level publications under the name *Nats Wolraf*, the mirror image of my first and last names, *Stan Farlow*. Although Nat's career as a purveyor of mathematical nuances never reached the stratosphere like those of fellow Mainer,

Stephen King, they did provide motivation for Nats to carry on. Once I got infected with that time-honored habit, I reached out to the text-book field, where no doubt, at this very moment there are legions of students out there poking about in one of my seminal texts on calculus, finite math and partial differential equations, no doubt using my name in vain. Did they actually think the answers to the problems came with the book?

But for those readers who are not into the more serious recesses of the Queen of the Sciences, this book might just tickle their fancy, and do a minuscule amount of educating in the process. And if not, there are others that might fit the bill, including the following [cough] bestsellers.

- Ten Reasons for Not Naming Your Cat Calculus, Thales House Press

- Mathematics: Ain't There an App for That? Thales House Press

- Partial Differential Equations, Dover Publications

- Paradoxes in Mathematics, Dover
  Publications

Jerry Farlow

ΦΣΞΛΘΩ

PROFESSOR PICKLE AND THE OMICRON AFFAIR

———

When the legendary Professor Pickle touched down at Duckworth Academy for boys, it was rumored that a conspiracy of ravens landed on the Commons. The Clock on the Old Tower chimed six bells. The ivy at the north end of Mangrove Hall turned brown and fell from the vines. And not least of all, students of mathematics prayed for divine intervention as all hope of horseplay in their upcoming Calculus 101 class fled like rats on a sinking ship.

During my first year at Duckworth, I was one of the unfortunate souls assigned to Professor Pickle's beginning calculus class. I still recall the dire expressions on the faces of my fellow greenhorns who were internally bawling and thinking to themselves, "All hope abandon ye who enter here."

My old friend Crowder was so scared he was shaking like a wounded gazelle. He knew he was digging his own grave, and true to his word, he never made it through.

Although the vast majority of students in the class floundered in the wake of Pickle's endless barrage of arcane symbols, he was a mathematical utopia for the mathematical aristocracy, who were on speaking terms with Pickle's byzantine vocabulary, mopping up his every premise and inference. Professor Pickle was a mathematical prophet spreading the Word to fertile minds.

"Ah, mathematics, it's the poetry of logical

ideas," they cheered. "It's the most beautiful creation of the human spirit."

For us *hoi polloi*, however, drowning in an ocean of unintelligible babble, the human spirit took a holiday.

It generally took Pickle ten minutes to fill all the panels of the blackboard with equations, after which he returned to the starting line and began refilling them with replacements. I wrote like the wind, but could never begin to keep ahead of him as he'd come up behind me, lap me, and disappear down the board.

Then, all of a sudden and a thousand equations later, he would stop abruptly, stand back from the board, point to a rather odious-looking equation, and ask himself,

"Hmmmmmmmm, now let's see what we have here?"

By now, the class was scattered all over the board taking copious notes, but upon hearing Pickle's hallmark query, everyone would abruptly freeze and look in a befuddled manner at Pickle's prize equation. After a few seconds, however, every-one's look of befuddlement changed to horror as they knew that Pickle's question wasn't always *rhetorical*, and there was the distinct possibility he might ask one of *them*, what do we have here?

"Well, mister —- " was the dreaded phrase each member of the class feared as no one dared look him straight in the face. One could hear the shuffling of papers and the scratching of a pen-cil, like a rat scratching for grain, by some poor retch in a pathetic attempt to say why he shouldn't be interrupted from busy note-taking.

Then, after Pickle grilled some helpless soul within an inch of his academic life, he would reload his chalk and race off down the board, leaving the class wallowing in his wake. Then, ten minutes later and another million equations, as if by clockwork, he would stop again and ask,

"Hmmmmmmm, now let's see what we have here?"

If we hadn't a clue what was there before, I always wondered how he expected us to know what was there now? Of course, I wasn't about to raise the issue now, especially when I was so busy shuffling my papers.

I estimated that over the entire semester Pickle used his trademark phrase a thousand times. I actually saw the object of Pickle's famous question one time, giving me a semester's average of 0.001.

However, as much as Pickle is remembered at Duckworth for his Teutonic manner and famous utterance, he will always be remembered by what is known as the *Omicron Affair*.

The first thing you should know about Pickle was how he loved mathematical symbols. Generally somewhere in his first sentence would be the phrase

given an $\varepsilon > 0$ there exists a $\delta > 0$ such that ...

Before long the blackboard was filled with hundreds of equations that looked more or less like

$$\frac{\partial^2 \Psi}{\partial \theta^2} = \frac{1}{\varphi} \left\{ \lambda + \frac{2^n \eta}{\zeta} \frac{\phi''}{\phi} \right\} + \frac{2\delta}{\Pi} \iiint_\Omega \Pi_\alpha \left(\sigma \right) d\sigma$$

During one of his more ambitious lectures, Pickle ran through his favorite Greek letters of α, β, γ and δ so on, eventually getting to some of the lesser used letters like λ, θ and ω. For the average math teacher, a few Greek letters suffice, maybe an occasional ξ, μ or ς when the discussion becomes a little verbose.

During this particular lecture, however, Pickle seemed to run through the entire Greek alphabet, but had five minutes of class time left. He paced back and forth in front of the board, mumbling to himself and scratching his head. It was the first time the class had ever seen Pickle stumped. A classmate carefully shot me a half-grin out the corner of his mouth.

Finally, Pickle lifted a finger in the air. "*Aha,*" he announced proudly. The entire class leaned forward, the first time actually awaiting the answer to one of his questions.

"*Omicron!*"

"Omicron?"

"*Omicron!*" Pickle repeated. "We haven't used *omicron!*"

My God, I thought to myself. No one ever uses omicron. You can't tell it from the letter 'o' or a zero. The man has completely devoured the Greek alphabet.

"We're seeing mathematical history being made here today," a classmate muttered out the side of his mouth.

Well, that's it. For those of you who have never taken a course in mathematics, you probably don't appreciate Pickle's accomplishment, but it just might be the only time in the history of mathematics that a mathematics teacher has used the entire Greek alphabet in a single lecture. I consider myself fortunate to have witnessed it firsthand.

I only wish I knew what it was all about.

ΠψΔΦθΛ

2

A SINCERE APOLOGY TO MY FELLOW MATH-TEAM NINJAS

Dear Math Team Ninjas:

I am too distraught and embarrassed to confront you in person so I am sending you this genuine letter of apology for single-handedly blowing our chances at Math Fest 2016 last week.

All your scorn coming my way is completely justified. Totally and unconditionally.

And all those things I said about Mary? Completely untrue. She is really extremely smart in math and there is no way I should have taken her spot on the team. I'm sure she would have gladly assumed the duties as new club member and performed them skillfully.

To each of you who invested so much time studying every niche and corner of math, I let you all down. I just wanted to fit in with the math guys.

When that judge asked me what "factoring out x" meant, I should have known it had nothing to do with the X-Factor TV show. I'm sure that answer dropped us down several points right there. Only now with my future in math destroyed, I realize the full extent of my crimes. Not knowing "Newton" was the guy who started calculus and "Newton" had nothing to do with "fig cookies" probably ended our chances right then and there.

To each of you who worked all year planning on a rematch with Malcolm Tech, I was our weak link Malcolm Tech was dreaming of.

It goes without saying I sincerely apologize for the off-handed juvenile crack I made when asked about that conjecture in graph theory called the "ASS CONJECTURE" It was totally uncalled for.

And my wry response to the question about abstract groups satisfying the "TITS ALTER-NATIVE" ... I could tell right then the judges were not amused.

And then there was my crude response to the question about measure theory and the great mathematician Norbert Wiener's definition of "WIENER MEASURE" I just wasn't thinking.

And I certainly have no excuse for my totally inappropriate double entendre related to the impossibility of vectors to be combed smoothly on the surface of a sphere, known as the

"HAIRY BALL THEOREM." My god, that answer even got around to my own mother!

Only now, looking back, do I realize the damage that I did to our team. You all have the right to shun me.

I have a lot of making up to do if I ever want to apply for membership again. I am hoping that the APPLE PIE I sent you with the crust shaped in the form of a PI will go a long way.

I hope we can still be friends although I suspect I won't be invited to the PI DAY celebration you are hosting next March 14 ... even though I make a killer chocolate cake with the 3.14159... in frosting on the top. Too soon? Well, maybe the year after next?

Good luck next year at MATH FEST 2017. I'm sure the judges will have forgotten all about me. At least some of them.

Harvey P. Mackleby

Ex-Ninja Member

3

PANNING FOR GOLD IN THE HEAT EQUATION

———

I recently asked my lawyer friend Ziggy, DMV, WTF and HDTV if I could patent a mathematical discovery.

"No way," he said. "All mathematical discoveries belong to the public domain. If you could patent a mathematical discovery, it would open a can of worms like you wouldn't believe."

However, Ziggy wasn't 100% correct. In 1897 an Edward J. Goodwin, M.D. from Indiana obtained a patent for his construction of a circle and square, each having the same area while using only a geometric compass and straight-edge. This ancient problem, called the Squaring of a Circle, is one of the three famous problems of Greek antiquity. In 1882 the 2000-year-old problem was proven to be impossible as a result of the German mathematician Ferdinand von Lindemann's proof that π is a transcendental number. Although a fee was required for anyone using Dr. Goodwin's construction, being a patri-otic Hoosier, Dr. Goodwin did allow Indiana educational institutions to use his masterstroke for free. Sounds great except there was a slight fly in the ointment. His proof depended on the fact that the ratio of the circumference of a circle to its diameter is

$$\frac{4}{(5/4)} = \frac{16}{5} = 3.2$$

which is equivalent to $\pi = 3.2$, an approxima-tion of π less accurate than the 4000 year-old Babylonian estimate of $\pi = 3.125$.

Nevertheless, that minor bobble did not stop Dr. Goodwin, He convinced the Indiana General Assembly to introduce HR 246, known as the *Pi Bill*, to adopt his bill as Indiana's contribution to education. Not understanding the proof (and probably not reading it) the bill passed unanimously by a vote of 67-0. Unfortunately for Dr. Goodwin when the bill reached the Indiana Senate there was a mathematician present who informed the lawmakers the folly of their ways, thus ending Indiana's strange brush with the most famous number (and ridiculous law) in mathematics.

———¤¤¤ΞΞΞ¤¤¤———

Dr. Goodwin, aside, if ever there *was* a law that paid royalties to heirs of past mathematicians for past (legitimate) mathematical discoveries, it would make every solicitor on the planet think he or she had died and gone to barrister heaven. Imagine the following scenario in the First District Court of New York.

Court Docket: Sir Isaac Newton versus People of the United States of America.

Mr. Newton's lawyer is about to speak.

"Your Honor, on behalf of the descendants of Mr. Isaac Newton, formerly of Woolsthorpe, England, we hereby make the claim of 355 quadrillion dollars against the United States of America for unauthorized use of the Integral and Differential Calculus and from applications derived therein. We have shown that monies accrued by people of the United States over the past 350 years far exceeds the requested amount."

The lawyer contended the amount was not unreasonable. The airline industry alone could not get their planes off the ground unless they had the luxury of differential calculus, discovered by Mr. Newton in 1665.

This class action suit filed by the 2,534 descendants of Sir Isaac Newton is only one of several suits filed by descendants of famous mathematicians. An anti-infringement suit filed by a great, great, great, great, great, great, granddaughter of Simon

Pierre Laplace against General Motors for 'unauthorized and flagrant use of the Laplace Transform' got underway last week.

A witness for General Motors testified the car company had never used the Laplace Transform, but preferred the Fourier Transform instead. Unfortunately for General Motors, a lawyer for the Fourier family was in the audience and began to initiate court proceedings.

An even more bizarre case surfaced recently in Rome where the descendants of Girolamo Cardano sued the Monte Carlo Casino for 325 sextillion francs for its unauthorized use of the principles of probability. In a recent interview a lawyer for the Cardano family made the following argument.

> *"Casinos have been figuring odds for ages using principles set down by Cardano."*

As large as these settlements have been, they are peanuts compared with the mammoth suit filed by the heirs of great English logician and philosopher, Bertrand Russell. In the case of *Russell versus the World*, Russell's descendants are suing everyone for use of logic and common sense.

"Everyone owes us," one of his descendants said recently.

On the other hand, the unfortunate remark made by the great English pure mathematician, G. H. Hardy, to the effect that 'nothing I have ever done has ever had any practical value' has come back to haunt his heirs. So far they haven't collected a dime.

ΠψΔΦθΛ

Epilogue: The ancient problem of the *Squaring of a Circle*, which consists of constructing a square and circle with the same area using only the classroom tools of a compass and straight-edge, is one of the three famous problems of Greek antiquity. In 1882 the 2000-year-old problem was proven to be impossible as a result of the German mathematician Ferdinand von Lindemann's proof that π is a transcendental number.

4

TOO BEE OAR KNOT TWO BEE

Thank God for good 'ol Silicon Valley. Just when I thought I was getting left behind on the Information Highway, along comes a software package from Micro-Math Systems which has catapulted me to the front of the line.

You see, back in the days when my classmates were deep in the theories of Euler, Lagrange, and

Plateau, I was at the movies watching Larry, Curly and Mo. But thank goodness, now with Micro-Math System's new Android App, Math-whiz, I have become a math whiz myself for only $39.

Mathwhiz is an App that accepts mathematical gobble-de-gook and turns it into some of the most sophisticated mathematics this side of MIT. And did it work! After entering my disgraceful little equation

$$x + 1/y - z = \text{cat}$$

Mathwhiz found my myriad of errors and reproduced it correctly as

$$u_t = \alpha^2 \left(u_{rr} + \tfrac{1}{r} u_r + \tfrac{1}{r^2} u_{\theta\theta} \right) + F(r, \theta, t)$$

Another recent product out of Silicon Valley is WriterWhiz, an App for your iPad that accepts misspelled gibberish, dissects it, then turns out some of the best prose this side of Stratford-upon-Avon. After entering my repulsive little verse:

———¤¤¤☲☲☲¤¤¤———

Rozes are Red
Violuts are blu,
All da flowers in heavin
Are nut as sweaty as yu.

———¤¤¤☲☲☲¤¤¤———

WriterWhiz regurgitated it as:

———¤¤¤☲☲☲¤¤¤———

O, speak again, bright angel! For thou art
As glorious to this night, being o'er my head,
As is a winged messenger of heaven,
Unto the white, up-turned, wondering eyes
Of mortals that fall back to gaze on him
When he bestrides the lazy-puffing clouds
And sails upon the bosom of the air.

———¤¤¤☲☲☲¤¤¤———

The moral is don't sweat all those outmoded
subjects like reading, writing and arithmetic.
Today, computers do all that drudgery and
unless they break down, you won't ever haf no

trubel wif nothin you ever deside tu du. An thats a fack.

$$\Pi\psi\Delta\Phi\theta\Lambda$$

5

A 49-PROVERB ODYSSEY THROUGH MATHEMATICS

———————

Someone once famously said

"Why are numbers beautiful? It is the same as asking, why is Beethoven's Ninth Symphony beautiful? If you don't see why this is so, no one can tell you."

In fact, the quote was the passionate belief of the Hungarian mathematician Paul Erdõs, who was famous for his introspections of mathematics.

Another observation by one of the great minds of the Queen of Science was due to the Russian mathematician Nikolai Ivanovich Lobachevsky, who observed

> *"There is no branch of mathematics, however abstract, which may not some day be applied to phenomena of the real world."*

It can be argued that mathematics has allowed us to understand the world more than any other intellectual discipline.

Reading people's thoughts about mathematics provides a glimpse into the nature of the beast. Whether you agree or disagree, we pass off a few viewpoints from the massive pile of opinions that have been floating around the mathematical atmosphere for over 2500 years, possibly even before Thales of Miletus realized the importance of the mathematical proof.

—¤¤¤ΞΞ¤¤¤—

1. Mathematics is the last discipline that comes to a developing country. It is only after everything else is developed that people turn to abstract thought. Say what you will about mathematics, at least it is right.

2. Mathematics is the most democratic of all disciplines. No one can discriminate against the laws of logic.

3. Mathematics is like an oak tree. Great things come from small beginnings.

4. Religious differences breed wars, mathematical differences breed new ideas.

5. After all is said and done, mathematicians are deduction robots.

6. Mathematics and science are the yin and yang of knowledge. Mathematics is deduction and science is induction.

7. Mathematics is the keeper of infinity.

8. A good theorem is like sausage and laws. The finished product may be a delight, but you wouldn't want to see it being made.

9. In the long run, one can better defend a country by teaching its children mathematics than by building tanks.

10. Mathematics waters the imagination as spring rains water a budding flower.

11. In mathematics you don't understand things, you just get used to them.

12. Eventually mathematics will become a necessity and then we mathematicians will rule the world.

13. The brightest ideas of mathematics are just that until they are proven.

14 The test of an axiom lies in the theorems it produces.

15. A mathematician must believe in both the possibility of God and the Devil.

16. In deciding whether to become an engineer or a mathematician, one must decide whether one likes to do or to think.

17. To learn mathematics is to acquaint oneself with the best that mankind has to offer.

18. In mathematics, only the genius gets lucky.

19. The whole of mathematics is not comprehensible to any one person.

20. To anyone not instructed in mathematics, the world must seem like a wonderful place, much of which is hidden from view.

21. I would rather have a dozen root canals than to work a single word problem.

22. Say what you like, there is no demagoguery in mathematics.

23. A good mathematician is one who makes the smallest amount of ideas go a long way.

24. Mathematics is the art of drawing necessary conclusions from sufficient premises.

25. The work of a great mathematician will effect future generations, but even he knows not where his theorems will lead.

26. The problem that makes the study of integers so hard is that they are so simple. What we must do is make them complicated.

27. Mathematics, it's just one damn theorem after another.

28. A person who is both a mathematician and a poet has it all.

29. The difference between a mathematician and a politician is that a mathematician tries to say the most with the least number of words, a politician does the opposite.

30. If mathematics were dogmatic, there would be no mathematics.

31. Mathematicians make their own language so no one else can understand them.

32. The old is never destroyed in mathematics. It is the foundation for new ideas.

33. An axiom is intuition that has passed the test of time.

34. If mathematicians would only realize how much they bore everyone.

35. One way to end a conversation is to tell everyone you're a mathematician.

36. Mathematics is not interested in race, creed, or religion. It is interested only in mathematical truths.

37. Now really, just where would we be without the Greeks?

38. Mathematics is the highest rung of human thought.

39. In all reality mathematics is beyond our mental facilities to understand it.

40. How can a finite mind comprehend the infinity of mathematics?

41. Mathematicians are the soothsayers and witch doctors of the 21th century.

42. Mathematical history is in essence a history of great ideas.

43. You can tell where a nation is going by the mathematicians it produces.

44. Mathematics is simply ideas reduced to their ultimate essence.

45. Obvious is the most dangerous word in mathematics.

46. Mathematics from the right vantage point possesses not only truth but beauty.

47. Mathematics is music for the mind.

48. Mathematics may just well be the language of God.

49. Mathematics may just well be the language of God.

ΠΦΓΛΘΩ

6

LISTEN UP, I'M NOT GOING TO REPEAT THIS

I'm starting to get annoyed with some of you. One would think that after all those endless hours in beginning algebra, you would have the technique of "simplifying an algebraic equation" mastered by now. But no doubt there are some among you who just don't get it. So pay attention, For that reason, I would like to give a short tutorial on how to simplify a mathematical equation. I'm not going to repeat this so stay alert.

Suppose you want to simplify the basic algebraic expression

$$y = x - x^2$$

As some of you probably learned in your algebra class. it is already in simplest form, but you still have a lot to learn. A few of you smarter ones will no doubt factor the right-hand side as

$$y = x\left(1 - x\right)$$

and choose this for your answer. Well, *au contraire*. We all agree the two above equations are equivalent and that the second one might seem simpler to some, but it is *still* possible to go a little further if you use a little imagination.

We begin by taking the reciprocal of each side of the last equation, getting

$$\frac{1}{y} = \frac{1}{x(1-x)}$$

and if we carry out the division of the ratio 1/(1-x) which appears on the right-hand side, we get

$$\frac{1}{y} = \frac{1}{x(1-x)} = \frac{1}{x}\left(1 + x + x^2 + \cdots\right)$$

We can now multiply each side of this equation by the expression

$$\frac{e^{-x^2}\cosh(\pi x)}{\sqrt{1+x^2}}$$

and integrating from 0 to z, yields the resulting equation

$$\int_0^z K(x)\left(\frac{e^{-x^2}\cosh(\pi x)}{\sqrt{1+x^2}}\right) dx$$

where of course

$$K(x) = \frac{1}{x} + 1 + x + x^2 + \cdots$$

The above integral on the right is now easily recognizable as the *Kleinhopper Integral* and so by integrating again and solving for y, we get the final form

$$y = \iiint_\Omega \sum_{k=1}^{\infty} \sum_{j=k}^{\infty} \left(\frac{\partial^2}{\partial x \partial y} \left(A_{jk}^{\alpha\beta} \frac{\partial u}{\partial x} \right) \right) + \int_\Delta \left(\sum_{k=1}^{\infty} a_k e^{-(k\lambda x)^2 x} dx \right)$$

Of course, this equation could be written in oblate spherical coordinates, *but what would be the point?*

$$\Pi\psi\Delta\Phi\theta\Lambda$$

7

AN OLD PROFESSOR
YOU MIGHT RECALL

———

Some of us old sages were pontificating in the faculty lounge the other day when Professor Fiddleman, BS, WTF, and HDTV announced he had written a guidebook detailing the myriad of pedagogical tools he had accumulated over the years on the effective way of teaching mathematics. Later, I was privileged to obtain a copy of Professor Fiddleman's book and would like to share its salient contents with the reader. Possibly, you know Professor Fiddleman, or at least, had a Fiddleman in your past.

Fiddleman's Canons for Teaching Mathematics

- **Never Over-Prepare**

It is important that you never over-prepare a lecture. If a lesson is laid out in perfect logical order and the ideas presented too clearly and succinctly, it gives the impression the subject is lacking in content. It is advisable to ramble from topic to topic so the student is able to see the richness of the subject. If the student is able to understand everything that is being said, the student will believe subject can be mastered without help from the professor.

Also, at the end of the lecture, when students are rudely putting on their coats and performing other annoying gestures intended to let you know you've run ten minutes over, just say, "Now here's an interesting little theorem that I'll bet you've never seen before, it'll only take me a few minutes to prove." Introducing new material at this time has several useful benefits for the student. Some students have been known to develop the stamina of a marathon runner dashing hell-bent across campus trying desperately to make their next class.

————¤¤¤ΞΞΞ¤¤¤————

- **Effective Blackboard Techniques**

There are several rules every young professor should know when it comes to blackboard techniques. The first rule is to lecture to the blackboard. This will let the students know your fascination with the subject matter while at the same time muffling your voice so as to keep them on their toes.

It is also important that you stand in front of the equation you are writing so that the student can develop a vivid imagination. And, of course, immediately after you've finished writing the equation with your right hand, you should begin erasing it with your left. This technique provides dozens of useful lessons for the student. First of all, to keep up, the student must cultivate the note-taking skills of a courtroom stenographer. Then too, students in the back row, jousting back and forth trying to see around you, receive the benefit of an aerobics workout.

Fiddleman hard at work inspiring eager learners.

And, last but not least, it is important when writing on the blackboard, to write large enough for everyone to see; generally characters about a half of an inch high are more than adequate. If students don't have the foresight to sit in the front row and bring binoculars to class, they're probably not smart enough to understand your lecture anyway.

———¤¤¤☒☒☒¤¤¤———

• **Handing Back Exams**

Never, under any circumstance, should you return a student's graded exam until at least eight weeks after the exam is taken. This will allow adequate time for the student to forget all the stupid reasons about the exam being unfair, as well as allowing time for the student to forget how to work the problems, thereby sacrificing his or her ability to argue for more points.

———¤¤¤☒☒☒¤¤¤———

• **The Proper Professorial Image**

The image every professor should cultivate is that of the campus eccentric, walking into the wrong classroom and giving your advanced calculus lecture to a group of history students goes a long way. Also, talking to the blackboard after the students have left the classroom can also be effective.

It goes without saying you should never wash

your socks or comb your hair. If your hair gets in your face, just keep it in place with a paper clip.

• Office Hours

It is axiomatic that you should never be in your office during office hours, if for no other reason, it gives the student the impression you have nothing better to do with your time. If you do honor an office appointment, tell a student you will be in your office at 8 A.M. but appear no sooner than 11 A.M. You can tell the student, who will be camped in the hallway outside your office, you had to discuss important university matters with the dean.

• Discipline in the Classroom

It is important that a professor maintain proper discipline in the classroom so that academic decorum is upheld. As an educator with many years of experience, I have never had a single problem with classroom discipline. I talk politely to the students, give them respect, and take my Colt 45 with me and place it on the table at the front of the class. Generally, the class is so quite you can hear a pin drop.

————¤¤¤☲☲☲¤¤¤————

- **Effective Use of the Uh Word**

In order that your lectures show intellectual reflection and depth, you may want to sprinkle in several "uhs" here and there. For example, instead of solving an equation quickly and precisely, it is far better to say

"now when solving the, uh, Steinmetz equation, we, uh, take the, uh, left-hand side and, uh, then multiply, uh, the ... , uh, solution.

This will impress upon the student your contemplative nature and your ability to talk and think at the same time. And, of course, no one will understand a word you're saying, avoiding annoying questions after class.

———¤¤¤ΞΞΞ¤¤¤———

- Homework

Homework always plays an important role in any student's experience. It is by working assigned problems that the student learns. However, it is important that the assignments are not too long or too difficult. Generally, 30 to 40 problems each period is about right. Of course, there is no way you should be expected to grade that many so simply let your graduate assistant mark each one off with a plus or minus.

———¤¤¤ΞΞΞ¤¤¤———

- Notes Versus Books in the Classroom

Never force your students to buy an expensive textbook when you can furnish them with a set of your own timeless notes at only 95% of the cost of a new book. Handing out a three-inch-thick pile of notes has all kinds of benefits, if nothing more than giving the students a sense of camaraderie as they groan collectively while the notes are being handed out.

They also give the student a sense of adventure, knowing that half the equations contain typos, thereby teaching the students not to believe everything they read. Also, students will develop the reading skills of a classical linguist as they try to decipher the hundreds of words smeared by the copying machine.

Also, they give the student a sense of history, especially if you haven't reworked them in thirty years.

$$\Pi\psi\Delta\Phi\Theta\Lambda$$

8

HOLLYWOOD'S FUNNIEST (MATHEMATICAL) LINE

In the category of what's gut-busting funny to one person but cold-stone sober to another, I once gave a lecture about artificial intelligence to a group of government types. This was in Washington D.C. back in the 1960s when it was clear that computers and artificial intelligence were going to take over the world. Whether that prediction has come to pass I leave that question for wiser minds, but I had just read some enlightened opinions on the matter and was psyched.

My introduction consisted in explaining the various types of intelligence, starting with the highest level and working downward. I began with human intelligence and explained its various properties, such as its capacity for abstract thought and problem solving. I then decided to lighten the pedantic mood of the talk by telling the audience there were two levels of intelligence below the human variety, the next one being artificial intelligence, which was the topic of the talk, and last but not least, the lowest possible form of intelligence, chuckle, chuckle, being military intelligence.

I was a little surprised that my quick-witted crack about military intelligence didn't harvest the expected chuckles from the audience. As a matter of fact the auditorium was so quiet you could hear the veritable pin drop. Possibly because, unbeknownst to me, 99% of the audience consisted of non-uniformed army and navy personnel from nearby government installations.

Every profession, including mathematics, has its own "lawyer jokes," which are generally characterized by being bad. If I had a penny for every dumb math joke I've ever heard, I'd put them in a sack and beat the next person to death who tried to tell me one. However, that said, there is *one* math joke any mathematician will laugh his or her head off, although, although I suspect army and navy personal might only stifle a smile. It goes like this:

————¤¤¤ΞΞΞ¤¤¤————

Two (male) mathematicians are in a bar and the first one says to the second one that the average person knows nothing about mathematics, whereupon the other mathematician disagrees and says that the general public is more familiar with mathematics than one thinks.

The first mathematician then goes to the washroom and in his absence the second one calls the attention of a waitress and tells her in a few minutes when his friend returns he will call her over and ask her a question, whereupon all she has to do is say one-half x squared.

She repeats "one-half, uh, er, x sued'"

He repeats "one-half x *squared*."

Finally, after some mumbling, she manages to say the response correctly.

The first mathematician then returns and the second one proposes a bet to prove that most people know something about math. He says he will ask a waitress to state the antiderivative of x. The other mathematician laughs but agrees. The second mathematician then calls over the waitress and asks "what is the antiderivative of x?" The waitress answers without hesitation "one-half x squared" and as she walks away says over her shoulder "*plus an arbitrary constant.*"

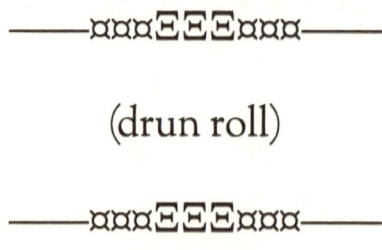

(drun roll)

Well, there you have it. Don't laugh too hard.

But give me another chance. For redemption, I will now relate the funniest line about mathematics ever uttered in a Hollywood film. You won't be disappointed.

Hollywood has always had a split personality when it comes to mathematicians. On one hand, they are portrayed as otherworldly think-tanks, able to scribble down an equation and then solve the thorniest of the world's problems. On the other hand then are depicted as, well, a little slovenly, awkward loners devoid of social skills, and goes without saying not overly popular with the ladies ... and of course, usually male. Other than independent documentaries about Sofie Germain and Hypatia, I know of no Hollywood movie that centers about a genius woman mathematician.

When I was in college back in the early Pleistocene Era, I decided to relax and take in a flick. I don't remember anything about the movie except one memorable line, a line so outrageously hilarious that when I finally managed to stop convulsing with laughter, the theater was so quiet you could hear a pin drop. I slunk down in my seat to avoid the cruel stares coming from everyone around me.

Like I said, I don't remember anything about the movie, except for one thing: There was a rather "square" (as they used to say back in the '50s) high-school algebra teacher, a street-wise, high school "hood" (as they used to say back in the '50s), and, last but not least, his older and, ..er, "working" sister (as they've said since the beginning of time).

Anyway, the hood was flunking the math teacher's algebra class (if I remember, he had to pass to get accepted into the Hell's Angels), so to get a passing grade, he asks his older sister (believe me, she was no rocket scientist herself) for help. Well anyway, the sister decides to use all the tools she has available (sexist innuendo aside, they were considerable), which sets the stage for the famous "math scene."

The algebra teacher, which Hollywood pulled out all stops in portraying him as somewhere between very socially inept and extremely socially inept, is having his nightly blue-plate special at the local hash house when in comes the older sister, who proceeds to slink across the room towards him. Well, Mamma Mia, talk about spooking the natives, the algebra teacher takes one peek over his horned-rim glasses and starts drooling like a basset hound.

She then stops directly in front of him, one hand positioned strategically on a hip, and leans over, and I don't mean she leans over, I mean she leans waaaaaaaaay over. Finally, after a few minutes in which she allows his pulse to push past a couple of hundred, she says, doing a damn good impression of Mae West *"What'd ya know about x, honey?"*

Well, the algebra teacher almost dislodges an eyeball, trying to maintain eye contact, all the while trying to cop a leer.

Finally, however, he regains full control, looks her straight in the eyes and says in a quavering but authoritative voice that would make any mathematician proud,

"*Ma'am, are you referring to the algebraic x?*"

— Pause for laughter —

Well, that's it, that's the line, "*Ma'am, are you referring to the algebraic x?*" I know, for most people it doesn't rank up there on the chuckle meter with the great funny lines of cinema history, but for me it was the best.

I only wish I remembered the sister's response.

ΠψΔΦθΛ

9

IN GOD WE TRUST: ALL ELSE REQUIRES PROOF

It is often said there is no relation between mathematics and religion. It is also said that mathematics and religion are directly related. The truth is that there are aspects of mathematics and religion that are related and aspects that are different.

On the similarity side, both mathematics and

religion are based on assumed truths. In mathematics, the assumed truths are called axioms or premises, whereas in religion they take the form of faith in a supreme deity and canons for moral conduct. But from this point on cracks between the disciplines begin to show. In mathematics the axioms along with accepted rules of logic are used to deduce new mathematical understanding, called theorems. In religion, the concept of a logical proof is foreign. It is the faith in a supreme being and its moral teachings that translate into righteous principles.

Although comparing mathematics and religion is probably a boring exercise to most people today, it was not the case in past times.

Pythagoreans

There was a time in history when mathematics and spiritual ideology merged their paradigms for the purpose of understanding. The Pythagoreans was a bizarre religious sect, created abound 530 B.C. by the Greek mathematician Pythagoras of Samos.

God was a mathematician to the Pythagoreans and Number was their First Commandment as manifested in their logo, "All is Number," or "God is Number," engraved in stone at the entrance to the Pythagorean School.

The holiest of all numbers was the "tetractys" or ten, which they represented geometrically as an equilateral triangle, consisting of 10 points arranged in four rows.

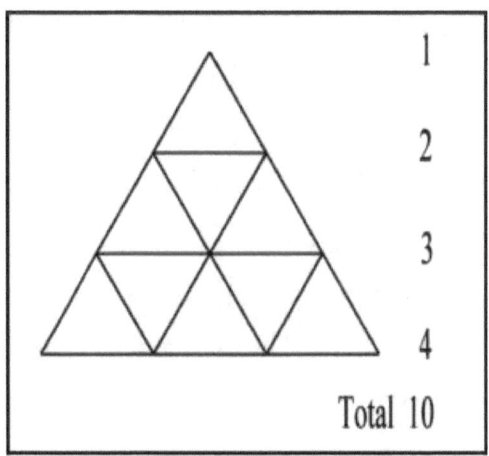

The Pythagorean tetractys

The Pythagoreans are credited for their contribution of introducing a more rigorous mathematics from what went before, building understanding from first principles or axioms, using logic to deduce new ideas.

The historian, Carl Boyer, once noted:

> *Never before or since has there been the unique interplay of mathematics and religion in the ideas developed by the Pythagoreans.*

The Pythagorean's religious passion about numbers is exemplified by the legendary account of the Pythagorean, Hippasus of Metapontum, who was thrown overboard off the coast of Greece for his unmentionable sin of proving that the square root of two is an irrational number (i.e. not a fraction). This fact was contrary to the Pythagorean's religious obsession with whole numbers and their ratios, and the mere thought of a number not being whole or a fraction, was considered heretical.

Infinity

If there is one unifying bond connecting mathematics and religion, it is the concept of infinity.

In mathematics, we say the numbers 1,2,3,... constitute an infinite collection of objects, while in religion "infinite" God refers to God without limits, endless, immeasurable, unconfined, omnipresent, ... But it is only in mathematics where the concept of infinity is brought to a more complete resolution through the research of mathematicians like Georg Cantor, Richard Dedekind, and others.

Nevertheless, many mathematicians, such as greats as Hermann Weil, Sir Arthur Eddington, and Nikolai Nikolaevich Luzin say that religious thought has influenced their mathematical thinking,

The German mathematician Hermann Weyl observed the spiritual side of mathematics when he wrote:

> ... *pure mathematical inquiry by its special character, its certainty and stringency, lifts the human mind into closer proximity with the divine than is attainable through any other medium.*

The mathematician J. D. Barrow goes even farther when he connects G?del's incompleteness theorem (roughly states that in any branch of mathematics there will always be propositions that can't be proven true or false) with religion:

> *"If a religion is defined to be a system that contains unprovable statements, then G?del has taught us that, not only is mathematics a religion, it is the only religion that can prove itself to be so."*

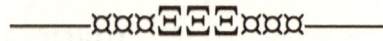

Immortality

One characteristic, embraced by most religions, which is also present in mathematics, is the concept of immortality. In religion it's the immortality of the soul, in mathematics it's the immortality of mathematical laws. The algebraic law

$$(a + b)^2 = a^2 + 2ab + b^2$$

was true a thousand years ago, is true today, and will be true a thousand years hence.

Throughout the ages, pious and atheist mathematicians alike have tried to prove and disprove the existence of God and immortality. For them, it was only natural that humankind's most intellectual activity should be applied in proving or disproving humankind's most sacred divinity.

During the Reformation in Europe, some of the greatest mathematical minds lent their talents to this activity. Although some mathematicians were devout Catholics, others were resolute anti-Catholics, yet others were against all forms of religion. Their commonality, naturally, was the application of mathematics in the support of their own convictions.

The most famous proof of God's existence was performed by the great 18th-century Swiss mathematician Leonhard Euler (oi'ler). Euler was deeply religious and anti-Catholic and used his reputation as an eminent mathematician to humble those whose beliefs were at variance from his own. He once humbled the atheistic French philosopher Denis Diderot, when both Euler and Diderot were guests of the Empress Catherine of Russia. The story has been described by the English mathematician August

DeMorgan in his 1878 book, *Budget of Paradoxes.*
He writes:

——————¤¤¤ΞΞΞ¤¤¤——————

... Diderot paid a visit to the Russian court at the invitation of the Empress. He conversed very freely, and gave the younger members of the court circle a good deal of lively atheism. The Empress was much amused, but some of her consolers suggested that it might be desirable to check these expositions of doctrine. The Empress did not like to put a direct muzzle on her guest's tongue, so the following plot was contrived. Diderot was informed that a learned mathematician was in possession of an algebraic demonstration of the existence of God, and would give it to him before all the court, if he desired to hear it. Diderot gladly consented. Though the name of the mathematician was not given, it was Euler. He advanced toward Diderot, and said gravely and in a tone of perfect conviction: Monsieur, , donc Dieu exist; repondez! Diderot, to whom algebra was Hebrew, was embarrassed and disconcerted, while peals of laughter rose on all sides. He asked permission to return to France at once, which was granted.

———¤¤¤ΞΞΞ¤¤¤———

The actual truth of the episode has never been taken seriously by math historians, but it makes for a good story, which of course, is at least as accurate as Euler's ridiculous proof.

ΠψΔΦθΛ

MATHEMATICS IN
IAMBIC PENTAMETER

———

Have you ever noticed the way poets have always avoided the subject of trigonometric functions? My own theory is that the only word that rhymes with "cosine" is "bovine" and who wants to read a poem about cows, except possibly the Ogden Nash classic:

———¤¤¤ΞΞΞ¤¤¤———

The Cow

The cow is of the bovine ilk,

One end is moo, the other milk.

It's safe to say there wasn't a rhyme Odgen Nash didn't like. From his brief to-the-point poem about fleas:

Adam, had'em

to the practical

If called by a panther, don't anther

the man was a rhyming dictionary.

Although we hope the readers of this magnum opus have a more cultivated taste in poetry than what some *littérateurs* refer to as poetic buffoonery, we nevertheless offer the reader a potpourri of poetic whimsy, Odgen Nash style. And if that's not enough, they're about mathematics! Enjoy!

Ode to Arthur Cayley

I greatly admire Arthur Cayley,

He proved a theorem almost daily.

——¤¤¤⊟⊟⊟¤¤¤——

Solution of Fermat's Last Problem

An Englishman named Wiles discovered the key,

To Fermat's Last Problem using geometry.

By proving the sum of two powers,

Is an integer to the power,

If and only if the power is smaller than three.

——¤¤¤⊟⊟⊟¤¤¤——

The Right Triangle

A right triangle, you can easily deduce,

Has a right angle, two legs, and a hypotenuse.

But are you aware,

That the hypotenuse square,

Is the sum of squares of the legs.

QEDuese.

————¤¤¤ΞΞΞ¤¤¤————

Professor Tician

Behold a famous mathematician,

A name known to all as Professor Tician.

The Dean exclaimed "No one surpasses!"

And put him in front of five algebra classes.

One day he missed ten equations straight,

He had, the students informed him later,

Forgot fifteen terms and a monomial.

Professor Tician could but smile,

"You mean," he said, a binomial."

————¤¤¤ΞΞΞ¤¤¤————

Tribute to Sir Isaac Newton

Here's to Newton, Isaac Newton,

Rootin tootin, Isaac Newton.

He was most astuten, there's no disputen,

*So let's all give a saluten to Sir Isaac
Newton.*

Yes, he lacked haute couture and social suavity,

But he gave us optics and the Law of Gravity

And yes he was moody and a little bit privative,

But he balanced all that with integral calculus

and the first derivative.

So let's raise our glasses and give a toast,

To the man who taught the world the most.

For who among you will refuten,

The lasting genius of Sir Isaac Newton.

———¤¤¤ΞΞΞ¤¤¤———

The Monomial

The monomial has a single term,

The binomial has two.

Or is it the other way around?

I'm never sure. Are you ?

———¤¤¤ΞΞΞ¤¤¤———

———¤¤¤ΞΞΞ¤¤¤———

Professor Snarf's Calculus Exam

Professor Snarf gave a calculus exam today,

He said it would be as simple as pie.

But when he told us to begin,

I thought that I would die.

He trotted out limits, he dangled integration,

He demanded the solution to Laplace's equation.

A Riemann integral, a Riemann sum,

*There were more limits on this test than in all
Christendom*

Ratio tests, M-tests, pap-tests, and infinity,

Followed for what might be the Holy Trinity.

And then I soiled my BVDs,

When he asked about vector analyses.

And then he shook up my cerebral architecture,

*When he asked for a proof of the Riemann
Conjecture.*

*Then finally something about the Hopf
bifurcation,*

And oh God, another Laplace's equation.

But Professor Snarf said he'd scale our grades,

And told us not to cry.

But when he returned my test to me,

I thought that I would die.

He exposed my surds, he refuted my queries,

He even poked holes in my infinite series.

He knocked down my limits and then he would gloat,

He even took umbrage with my asymptote.

He said my theorems did not deduce,

And blithely crossed off my hypotenuse.

And once he got going he never looked back,

Until my exam was covered in black.

And finally he nixed my line integration,

And, oh God, he even marked off my Laplace's equation.

———¤¤¤ΞΞΞ¤¤¤———

———¤¤¤ΞΞΞ¤¤¤———

———

The Origin of Mathematics

Some say the Babylonians started it all when they
realized there was more to life than growing
tomatoes, cabbages, and cucumbers.

So they started scratching out wedge-shaped
symbols on little clay tablets, which eventually
turned out to be our present-day numbers

Yes, some will argue mathematics began along the
Tigris and Euphrates in old Babylonia,

Although there are those who argue it began with
Thales of Miletus along coastal Ionia.

Now others will say mathematics began with
Pythagoras, Archimedes and a few other Greeks,

Although more than one Hindu will say it began
with the Sikhs.

But wouldn't you agree it's rather puerile.

If we didn't at least consider the Nile

And who would ever want to displease,

Wu Wang, Huang-tese and a billion Chinese.

But I think if the history of math were accurately told,

Of the first man and woman who spun numbers from gold,

We'd have to go back past Greece and the Nile.

Beyond India and China by a country mile.

Past Moses, Noah, and the Queen of Sheeba.

$$\Pi\psi\Delta\Phi\Theta\Lambda$$

HOW I (ALMOST) PROVED FERMAT'S LAST THEOREM

At the time I thought it rather strange. After all, any alien space jockey worth his plasma knows he's supposed to touch down in an alligator swamp in South Georgia and scare the holy be-jesus out of some Georgia hillbillies. This time, however, they decided to land in the Maine woods and scare the holy be-jesus out of us Maine *woodsbillies*.

I was taking an evening stroll behind my house when suddenly I saw it, the unearthly green iridescent glow through the trees, the oval-shaped saucer hovering above the clearing. I hid behind some bushes and watched as a small opening materialized on the underbelly of the capsule and several otherworldly creatures, each no more than three-feet tall, emerged. Each had a human-like frame but with unmistakably alien features, a rubber-like skin that stretched over a bony scaffolding and two huge eyes that radiated an eerie green glow from a pasty-white face. I watched as no less than a dozen of these creatures made their way down a ladder. They all appeared similar except for one, which had a row of medals on its chest. I took it to be their leader.

After they reached the ground, the leader looked in my direction and shouted, "Farlow, we're here to make a deal." My body froze. "What's wrong, Farlow, is this the first time you've ever seen *Orkians* before?" the leader said as the creatures surrounded me. One of the creatures waved his bony fingers before my eyes and suddenly my fear was gone. "That's better," the leader said. "Now let's get down to business."

"Business?" I stammered.

"You're a mathematician, aren't you Farlow?" the creature said. "And by the way, we don't like to be thought of as creatures, you know we can read minds, we prefer aliens. My name's Harry."

"Yes, Harry," I said. "I'm a mathematician."

"And not a very good one, we understand. This makes you the perfect candidate," the little creature said. A sharp pain ran through my body.

"I warned you, Farlow. We have ways for dealing with thoughts like those."

"I won't let it happen again, Harry," I moaned, "but I'm an excellent mathematician."

"Don't give us that," Harry said. "We know you haven't made a single discovery in your entire mathematical career."

"What about my proof of the *Widdlestein Conjecture*?" I protested.

"You call that a discovery?" Harry laughed. "Our kids learn that in kindergarten. But don't worry Farlow, we're going to give you the answer to the most famous mathematical problem in the world. We're going to make you a star."

"Wow!" I started to like the little guys.

"Just a few gallons of plasma should do," Harry said.

"What?" I asked.

"One discovery for just a little plasma," Harry said, "You didn't think we'd give you the answer to the most famous mathematical problem in the world for nothing, did you? A couple of gallons should do it right guys?" The other aliens nodded in unison.

"Wait," I protested. But before I could say anything the alien with the bony fingers waved them before my eyes.

————¤¤¤ЭЭЭ¤¤¤————

The next thing I knew I was stretched out on a metallic table looking up at a small alien in a white lab coat. He was shining a beam of light in my face with a strange unearthly object which, for lack of a better description, looked like an ordinary flashlight. I panicked and screamed out.

"What's wrong, Farlow?" Harry asked. "Haven't you ever seen a flashlight before. Stop playing around with that damn thing, Raymond. And take that doctor's coat off. Damn kids," Harry said.

My body lie enmeshed in a tangle of wires and tubes. Ooze of several degrees of *yeeeeuuuuck* flowed like sea water through the tubes. I noticed that one of the tubes seemed to end somewhere in an opening in my abdomen. "What's all the gunk in that tube?" I asked.

"Oh don't mind that," Harry said. "It's just something we add."

"What about our deal?" I protested, trying to get up. "You said you were going to make me a famous mathematician."

"Yeah, yeah," Harry said. He then crossed the room to a filing cabinet and removed a large manila envelope from the top drawer. "Here it is," Harry said. "The answer to your most famous mathematical problem, a problem you earth people have been trying to solve for 358 years, a proof of Fermat's Last Theorem."

"What?!" I yelled, struggling to get up. "I'm giving you my plasma for Fermat's Last Theorem? That problem was solved 20 years ago!"

"What?" Harry seemed surprised.

"That's right," I said. "An Englishman proved it."

"Oh, sorry about that," Harry said. "We've been on the road so long you know. When we left they told us you didn't have a clue. What is that old theorem anyway?" Harry asked.

"I thought you were so smart," I said sarcastically.

"It's been a long time since kindergarten," Harry said.

"Well, let's start with something on that level then," I said. I thought it best if I gave Harry a beginner's lesson on the problem.

"You'll agree that

$$3^2 + 4^2 = 5^2$$

don't you?" I asked.

"Our kids ... ."

"Yeah, yeah," I interrupted. "They learn it in kindergarten"

I also told Harry there are other pairs of numbers whose sum of squares is the square of a third number, such as

$$5^2 + 12^2 = 13^2$$

"What does all this have to do with Fermat's Last Theorem?" Harry asked, starting to get bored. I told Harry that although there are other integers like the above that satisfy the equation, the French mathematician, Pierre de Fermat, claimed there weren't any non-zero ones that satisfied

$$a^n + b^n = c^n$$

when the exponent n is greater than 2, like 3, 4, ... Fermat scribbled this claim in the margin of a book but said his proof would not fit in the margin. And so for 358 years many of the world's greatest mathematicians tried without success to prove Fermat's claim, known as Fermat's Last Theorem.

Finally, in 1994 a 40-year old English mathematician from Princeton University, Andrew Wiles, solved the problem.

"If you would have come 20 years ago, I'd been famous," I yelled at Harry. "*Harry, where the devil are you?*"

I then realized the aliens weren't the slightest interested in Fermat's Last Theorem, and were all clustered around a huge apparatus. Suddenly, I heard a voice cry out, *"How do you like the new me, Farlow?"* I looked up and saw the most hideous looking alien I'd ever seen, it was horrible. *I then realized I was looking in a mirror!*

Harry and some aliens looking for plasma

"*Aaaaaaaggggggghhhhhhhh,*" I screamed. WHAT HAVE YOU DONE TO ME?!" Looking around the room, I saw a dozen spitting images Brad Pitt.

"It's amazing what a little plasma will do for your complexion," one of the Brad Pitts said in a voice that sounded an awfully lot like Harry.

"Everyone wants to be a Brad Pitt, you'd think someone would pick a Clooney or a Redford."

"*Aaaaaagggggggggggggghhhhhhhhh*," I screamed out again. "No, no, ... give me back my face, keep the damn theorem ... ."

———¤¤¤ΞΞΞ¤¤¤———

".... wake up dear, you're dreaming again," someone yelled at me. I found myself sitting in my own bed, drenched in sweat. My wife was shaking me.

I looked at her but questions remained. Had I been dreaming, or had I actually been visited by creatures, OUCH, aliens? If so, might they return and give me the answer to the Riemann Hypothesis or what about the Goldbach Conjecture? Only time would tell. I could still be a famous mathematician.

"Proving Fermat's Last Theorem again dear?" my wife asked wryly.

"Of course not," I said. "It's been proven, go back to sleep.

$$\Pi\psi\Delta\Phi\theta\Lambda$$

12

WERE NUMBERS
PRESENT AT THE TIME
OF THE BIG BANG?

———

When Pythagoras of Samos scratched out his famous equation on a scrap of papyrus 2500 years ago, did he conclude

"Χολη Απαλλω, look what I've found,"

or

"Χολη Απαλλω, look what I've created."

Well, maybe no reference to Apollo, but he did raise the age-old philosophical conundrum of Platonism versus Intuitionism, or what most people are apt to call, how many angels can sit on the head of a pin?

Platonism

But, if you like to dabble in such enigmas, the (mathematical) Platonist philosophy holds the belief there is a mathematical world outside our world of space and time, where all mathematical truths reside. It is a world, independent of humankind, where mathematical entities and theorems dwell. Adherents of this philosophy are called mathematical Platonists, and when a mathematical Platonist proves a theorem or makes any kind of mathematical discovery, the person believes he or she has *discovered* (not created) a truth that *already in existence* in that distant world of mathematical truths. If Pythagoras had never discovered his famous theorem, it would still be out there in the cosmos of undiscovered mathematical entities.

The great Hungarian mathematician Paul Erdos (1913-1996) often referred to "The Book," where God keeps the proofs of mathematical theorems — a Platonist in the true sense of the word.

Intuitionism

On the other hand, mathematical intuitionism, introduced by the Dutch mathematician L.E.J. Brouwer (1881-1966) is based on the philosophy that mathematical ideas do not exist independent of humans, but are created as a result of human mental activity In other words, the Pythagorean Theorem did not exist before being created by Pythagoras.

The intuitionist view of mathematics has some far reaching implications when it relates to how mathematics is practiced. One consequence has to do with the *principle of the excluded middle,* one of the three Aristotelian laws of logic. The principle (or law) states that for any proposition, either the proposition is true or the proposition is not true, nothing else. Although few would question the validity of such a claim, for an intuitionist, the claim is not universally accepted, especially when the proposition in question relates to infinite sets.

The principle of the excluded middle is the basic tool used by mathematicians when proving theorems by contradiction. When proving the validity of a theorem or proposition by contradiction, one assumes the proposition is *not* true (i.e. false), then if one arrives at a contradiction of some type, one concludes one cannot assume the proposition false, and by appealing to the principle of the excluded middle, one concludes the proposition true. But intuitionists disallow the principle of the excluded middle, and so proofs by contradiction are considered invalid.

Another characteristic of intuitionism is its non acceptance of the infinite. Infinite sets, like the natural numbers [1, 2, 3, ...] or real numbers, are not considered valid mathematical entities since, unlike finite sets like [1, 2, 3] or [a, b, c], infinite sets cannot be constructed, only *imagined*. An intuitionist would say the natural numbers [1, 2, 3, ...] are "potentially infinite" since they are unending, but intuitionists do not consider the "completed set" as a legitimate mathematical entity.

Artists and composers consider themselves intuitionists (albeit non-mathematical) inasmuch as they imagine themselves creating their art and music from nothing, not discovering hidden scores and paintings already existing in some mythical world of art and music.

Although Platonism and intuitionism are distinct philosophies, people often have leanings for both philosophies. Imagine the philosophical dilemma of a sculptor staring at a block of marble. On one hand, the sculptor might think like a Platonist and imagine Michelangelo's *Pieta* is inside, waiting to be released by chopping off the stuff around it. On the other hand, the sculptor might view the block as an intuitionist, imagining the creation of a new work of art, which he will baptize, the *Pieta*.

The debate whether mathematical truths are discovered as a Platonist claims, or invented as an intuitionist claims, has been with us since the time of the Greek philosopher Plato, the originator of the Platonic philosophy. Today, if you do ask the average mathematician which philosophy he or she subscribes, they are apt to say they haven't given it a great deal of thought, but pressed will probably admit to being a Platonist. This belief is also shared by most of the great mathematicians of the past, including Newton, Gauss, Hardy, Cantor, and Godel.

ΠψΔΦθΛ

13

I'M NOT GONNA SUGAR COAT THIS: IT'S THE EXPONENTIAL, STUPID

―――――

I have to admit, I don't have the foggiest idea about fantasy football, bitcoins, or the importance of the Higg's Boson, but I do agree with physicist, Albert Bartlett, who once warned,

"Humanity's greatest weakness is its inability to understand the exponential function."

―――

The reason, of course, is that it grows faster than one imagines, sneaking up on us until ..."yippee" or "uh-oh." depending on the exponential growth of *what*. The simple paper-folding demonstration (don't do this at home) of folding a (big) piece of paper 0.01 inch thick on top of itself 50 times, resulting in a stack of paper 177,698,849 miles high, is just one example of exponential growth. If you're skeptical, get out your calculator and compute

$$\text{thickness} = \frac{0.01 \times 2^{50}}{12 \times 5,280} = 177,698,849 \text{ miles}$$

If you are wondering about how large a piece of paper you should use, the size of Wyoming will do quite nicely.

On a more realistic level, exponential growth is exhibited in the story of the Dutch explorers who in 1626 purchased the island of Manhattan from the *Canarsie Indians* for trinkets valued at $24. But before bestowing sympathy upon the Canarsies, recall the afore-mentioned words of Albert Bartlett, who warned of faulty conclusions from not respecting exponential growth.

In fact, from the point of view of the exponential function , one might argue it was the *Canarsies* who got the better end of the deal. Suppose the Canarsies were able to convert the gifts into $24 cold hard cash and deposit the amount in a bank that pays 6% interest. If the bank compounds interest annually, then every year the bank pays the depositor 6% the current principle, by simply multiplying the current principle by 1.06. Table 1 shows how an initial principle of $24 grows from 1626 to 2014 (388 years).

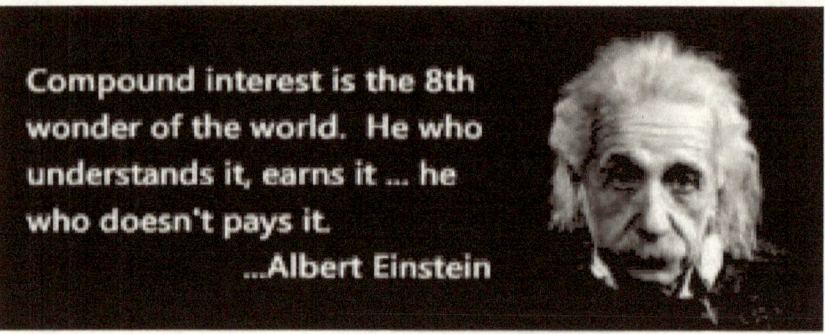

Compound interest is the 8th wonder of the world. He who understands it, earns it ... he who doesn't pays it.
...Albert Einstein

Year	Value of the Account
1626	$24
1627	$24 (1.06) = $24.44
1628	$24 (1.06)^2 = $26.97
1650	$24 (1.06)^{24} = $9717
1700	$24 (1.06)^{74} = $1,789.97
1800	$24 (1.06)^{174} = $607,339.88
1900	$24 (1.06)^{274} = $11,187,297.25
2000	$24 (1.06)^{374} =$69,920,552,399.62
2015	$24 (1.06)^{389} =$158,083,653,508.86

Table 1: Future of $24 at 6% compounded annually

Since banks compound interest continually, the future value of $24, deposited in 1626 at 6% interest, compounded continuously, 388 years later in 2014, would be

$$\$24\,e^{(0.06)(388)} = \$309,447,358,922.41$$

or roughly 309 billion dollars. It's a shame the Canarsies were not able to get 8% interest, else their account would have grown to the following value

$$2014\text{ value } = \$24\,e^{(0.08)(388)} = \$725,624,537,027,627$$

or 725 trillion dollars, no doubt more than the real-estate value of Manhattan today.

It just points out the old adage credited to Ben Franklin who once observed that "time is money," especially when talking about the exponential.

$$\Pi\psi\Delta\Phi\Theta\Lambda$$

14

GOD EQUATION: SENTRY AT THE PEARLY GATES

Wrong!" the old woman at the gate cackled fiendishly in a voice that could raise the dead, which considering the fact that everyone within earshot *was* dead, made the entire line jump to attention. The meek-looking man who was the object of her ridicule lowered his head and slunk off.

"Next," the old bag hissed. I had finally reached the front of the line, whereupon I approached the gatekeeper and gave her my manila envelope. She tore it open and began rifling through the papers. "Farlow?" she said looking over a pair of antiquated spectacles. "Is that *you*, little Jerry Farlow?

I looked into the beady eyes that peered from the face of this sorceress. Suddenly something inside clicked.

"Ms Hammerschnozle?" I asked. "Is that you?"

"Of course it's me. *But what are you doing here?*"

"We all die sometime, Ms. Hammerschnozle," I said. "I didn't want to go to the other place. But what are you doing here?"

"*Where did you think I'd be?*" she said snidely.

"Oh, I knew you'd be here," I lied through my teeth.

"But isn't Saint Peter supposed to be a *male* angel."

"Well, as usual Farlow, you thought wrong," The old bag said. "Times have changed. They finally got some religion up here and brought in some-one that knew a thing or two about math."

God! My ultimate worst nightmare! My infa-mous tenth-grade math teacher, Ms. Hammer-schnozle, whose endless aspersions on my aca-demic reputation almost a century before, was now a stand-in at the Pearly Gates for no other than, alas, Saint Peter. And she would determine whether my final destination would be the `ol harp farm in the sky or a permanent reservation along the River Styx.

"You know," I smiled at Ms. Hammerschnozle. "I've learned just a little mathematics myself since my school days. Why don't you just give me the test." I smiled smugly, knowing there was no way she could ask me something about math-ematics I didn't know.

"Ok, Farlow," she said. "What do you know about the *God Equation*."

"*The God Equation?*" I croaked. A river of sweat

the size of the Ganges ran down my face. "Uh, I don't think I'm familiar with that particular equation."

"It's God's own equation," she said. "This single equation contains the most famous numbers in all mathematics. If you can answer my questions about these numbers, you will be allowed to enter Heaven. If not you will … ."

"Yes, yes, I understand," I shuddered to think about the alternative.

"Are you ready, Farlow?" she asked. "The first number is about the number *one*. What do you know about 1?"

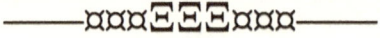

The Number 1

"Was she kidding?" I chuckled to myself. *Eins, Zwei* good `ol Numero Uno, the first of what we call the *natural numbers*

$$1, \ 2, \ 3, \ 4, \ ...$$

In addition to being the first natural number, it is the only number that has the property that the product is unchanged when multiplied by other numbers. For example

$$1 \times 6 = 6 \quad 1 \times 34 = 34 \quad 1 \times 68 = 68$$

"Ok," Hammerschnozle interrupted, "Enough of the kid stuff. I think we can get on to something more challenging." I waited for the guillotine to drop. "What can you tell me about π, the most famous number in geometry?"

The Number π

I chuckled to myself. Although π isn't an integer like 1, it is one of the most famous numbers in mathematics and was well known by Greek mathematicians over two thousand years ago. Quite simply, it is the ratio of the circumference of a circle to its diameter. The first eight digits of π are 3.1415926, which means that the circumference of a circle is slightly more than three times its diameter.

"Ok Farlow, you can stop with all that," Hammerschnozle interrupted. "You're halfway home. Are you ready for the third great number of mathematics?" she asked. "Each number gets just a bit harder."

I started to squirm a little. "Yes," I finally said.

"Ok," she said. "What can you tell me about the imaginary number i?"

The Imaginary Number i

I was now on a roll. The imaginary number i is one of the most fascinating numbers in all mathematics. You might say the origin of the number i began in the 16th century when Italian mathematicians Gerolamo Cardano and Niccolo Tartaglia tried to solve 3rd and 4th order polynomial equations. In the process they ran into the square root of -1, or $\sqrt{-1}$. They considered the square root of a negative number an impossibility and so it was taken as a 'non number,' but after some time mathematicians found it useful and so it was accepted as a number, although sometimes called an imaginary number, which we denote by i. The entire subject of complex numbers has applications in ...

"Ok, ok, you can stop now," Ms. Hammerschnozle interrupted. "It's clear you know a little about mathematics. You told me about the three famous numbers 1, π. and i. We now have only one more number to go but it is the most difficult. What can you tell me about the number e ?"

The Elusive Number e

"What?" I thought, no trick question? Every mathematician worth his white board knows about e. In some areas of mathematics, like differential equations, it is arguably the most famous number of all. The number e is a real number like the numbers 1 and π. However, it is more difficult to describe since it is generally defined as the limiting value of some other numbers. The Swiss mathematician, Leonhard Euler, defined *e* as the limiting value of the expression

$$(1 + 1/n)^n$$

as *n* gets larger and larger. It is used by engineers to describe growth and decay of ...

"That's enough," Hammerschnozle said at last, "No more sugar coating the questions."

Welcome
Mathematicians

Heavenly delights

Your final exam

I looked over her shoulder to see if the Pearly Gates had started to open. "Now just tell me about the God equation and we're finished," she said.

"*What?*" I said.

"Suppose we combine 1, π, *i and e* into the single quantity

$$e^{i\pi} + 1 = 0$$

she said. "What new number do we get?"

"*It's impossible,*" I yelled. Just the thought of combining these four numbers into the single number gave me chills.

"Wrong!" she barked. The value of $e^{i\pi} + 1$ can easily be found. *In fact it's zero!* The great Swiss mathematician Leonhard Euler used his famous equation

$$e^{ix} = \cos x + i \sin x$$

and simply plugging in $x = \pi$, he got

$$e^{i\pi} = \cos \pi + i \sin \pi = -1 + i\,(0) = -1$$

thus

$$e^{i\pi} + 1 = -1 + 1 = 0$$

"Up here we call it the *God Equation*," Hammerschnozle said.

"*What?*" I gasped in disbelief.

She continued, "This single equation brings together the five most famous numbers in all mathematics; the basic whole numbers 0 and 1 from arithmetic, the fundamental constant π of geometry, the complex number *i*, and the core constant *e* of calculus."

Next!" the old bag cackled heinously.

"*Aaaaaaaaaaaahhhhhhhhhhh*," I screamed as two burly angels grabbed me by my arms. "No, no," I protested. "I belong here. Give me another chance."

"*Wake up, wake up,*" I felt someone tugging on my arm. I was sitting up in bed, drenched in sweat, looking at my wife. "Those nasty little monkeys carrying you off again, Toto?" she said.

"Would you believe it? Did you know

$$e^{i\pi} + 1 = 0\ ?"$$

"Yeah, yeah" she said rolling over, "The God Equation. Go back to sleep."

$$\Pi\psi\Delta\Phi\theta\Lambda$$

15

DEATH, TAXES AND ANOTHER PROOF OF THE PARALLEL POSTULATE

I guess we'll just never know. Although some pointy-headed Egyptologists and archeologists tell us that the pyramids were built to house the tombs of the pharaohs, it is my own theory they were commissioned by a cult of ancient mathematicians to house the solution of the Riemann Hypothesis. You think I'm dancing with the fairies? Well maybe so, but there are a dozen other theories, including the one that Noah built the pyramids right after he finished off his arc, that would make my mine seem as logical as a proof in Euclid's Elements. .

Let's face it, there are those among us who march to a different tempo. In the late 1800s and early 1900s, there was a mathematical cult of "thank-God-we're-here-to-set-the-record-straight" individuals, whose goal it was to solve the three ancient problems of Greek antiquity — trisecting an angle, squaring a circle, and proving the parallel postulate. On the surface, this would appear to be a noble pursuit, if it weren't for the fact that the problems have been proven to *have* no solutions. It's like finding the smallest number greater than zero, there isn't any — whatever number you choice, someone can come along and take half your number.

But it's human nature to try to solve problems which others claim impossible, so these ancient problems of antiquity have attracted a mixed bag of mathematical Don Quixotes, who as their namesake before them, spent their days tilting at the proverbial windmill. But it wasn't that these individuals chased a nonexistent goal, what made their crusade so interesting. It was the fact some actually claimed success and shared their results with the general public in the form of self-published books. These books may not have been accurate, mathematically speaking, but they do make fascinating reading, being peppered with colorful boasts and personal "words of wisdom."

If you are unfamiliar with the three problems of antiquity, you are not alone. The problems are sufficiently obscure as to be known mostly to geometers and mathematical historians.

Trisection of an Angle

The first problem of antiquity, trisection of an angle, consists in trisecting an angle using only the grade-school tools of a compass and straightedge. Teachers during the 19th century often assigned this problem to their students if for no other reason than a lesson in humility. The general trisection problem was proven impossible by the French mathematician Pierre Wanzel in 1837. Wanzel proved that trisecting an angle was equivalent to solving certain cubic equations, which he showed cannot be solved using the straight-edge and compass method, thus most angles cannot be trisected.

Squaring the Circle

The second problem of antiquity, squaring the circle, consists in using the same two grade-school tools of compass and straightedge to construct a square with the same area as a given circle. This task was proven to be impossible in 1882 as a consequence that the number π was proven to be a transcendental number. (i.e. not the solution of a polynomial equation with integer coefficients)

Euclid's Parallel Postulate

The third problem of antiquity consists in proving what is known as Euclid's parallel postulate. If you can remember back to your high-school geometry days, you may recall that plane geometry was based on five postulates (or assumptions) and that the fifth postulate, which was kinda the most complicated, stated that two parallel lines never meet. What your teacher didn't tell you was that the early Greek geometers had the funny feeling that this fifth postulate could be proven from the first four postulates, thus making it redundant. So for 1500 years, mathematicians tried to prove Euclid's fifth postulate from the other four.

To make a long story short, there was no success until the 19th century when a Russian mathematician, Nicolai Lobachevsky, proved to the embarrassment of all, that Euclid's fifth postulate was *independent* from the first four postulates and thus cannot be proven either true or false. So in a sense mathematicians had been trying for fifteen hundred years to solve a problem that could not be solved.

———¤¤¤ΞΞΞ¤¤¤———

Given that the three problems of Greek antiquity have no solutions, it is intriguing that in the late 1800s and early 1900s several people sought answers to these problems — and even more intriguing that several people *found* answers. Here are some inspiring nuggets from publications of a few individuals who claimed to have mastered the unsolvable.

———¤¤¤ΞΞΞ¤¤¤———

The Man Who Made No Errors

Important Discoveries in Plane Geometry by P. D. Woodlock (Stephans Publishing Co.) Columbia, Mo. 1912.

In a 39-page pamphlet, a Mr. P. D. Woodlock claims to have solved the ancient problem of trisecting an angle. Mr. Woodlock dispels all doubts the reader might harbor when he says there are absolutely no errors in the book. He invites the reader to read the book "carefully and without bias." To set his work in proper framework, he writes,

> "This problem [trisecting the angle] has occupied the attention of geometers in all ages since the introduction of geometry as a science, yet all attempts at the problem have failed until now. That the author has been rewarded with the discovery of the true solution he confidently leaves to the consideration and judgment of geometricians throughout the world."

After this statement, he takes off with an elaborate series of diagrams, pictures, and definitions leading eventually to the strange concept of 'equalizing curved and straight lines.' He then reveals how he has managed to square the circle.

One hesitates to read the entire manuscript since its basic premise is the incorrect statement that $\pi = 22/7$.

The Philosopher Mathematician

Philosophic Reviews by Lawrence S. Benson (J. S. Burnton Publishing Co.) 1875.

Mr. Benson chides 19th-century mathematicians for giving up on the squaring of the circle when he writes:

> *"From the earliest accounts of mathematical science, mathematicians have given the most assiduous application to discover the solution to the squaring of the circle, with the invariable result of ill-success and despair. Finally these geometers have assumed a kind of indifference and pretend to ignore all attempts for it."*

Mr. Benson also has some qualms with calculus, where he finds flaws in the concepts of the limit, infinity, the derivative and the integral. He feels these ideas have no use except for straight lines — which, of course is exactly when they are *not* needed.

The Story of John A. ("I did it my way") Parker

The Squaring of the Circle by John A. Parker (S. W. Benedict, NY) 1851.

Publishing this volume in 1851, Mr. Parker found it necessary to hold back sales to a select few. He writes:

> "...is not published for sale, and no copy of it will be sold without the author's consent, at any price, as the author claims other privileges than a simple copyright, in respect to the alleged discovery."

Mr. Parker's general dislike and distrust of academia comes out when he says:

> "The professors of the schools, unable to demonstrate the truth, and unwilling to acknowledge their deficiencies, have thought it necessary to explain away their error, and in doing so, they have taken the opposite ground, and concluded, that what they have been unable to attain, is unattainable by others".

An example of the sort of mathematical reasoning that was withheld from the public is provided in one of the early pages:

"The circumference of a circle is a line outside of the circle thoroughly enclosing it".

The "Greatest Work Ever Written"

Atheism and Arithmetic by H. Hastings, Boston, Mass. 1885.

In a strange little lucubration that starts off humbly enough by presenting a rather incoherent discourse of the existence of God, a Mr. H. L. Hastings from Boston guides one on a lively tour of geometry and faith, religion and chemistry, arithmetic and the Celestial Time-Keeper, and last but not least, a somewhat curious concept known as "phyllotaxis." Lest any critic question the rigor of his deductions, an alternate proof of the existence of God is given in the appendix using the simple arithmetic principle of the *law of sevens*.

The Common Sense Theory of Geometry

The Common Sense Theory of Space by John Newton Lyle (privately printed) 1890.

Back to basics might have been Mr. Lyle's motto. John Newton "show me" Lyle rejected Lobachevsky's proof of the independence of Euclid's parallel postulate, stating:

> *"There is but one space, and that is a continuously extended, precisely alike everywhere, inflexible, and unbounded."*

Many of the great mathematicians and thinkers of the day are justly chastised here: Kant, Fichte, Helmholtz, and of course, Lobachevsky, all get their just due. He states of the great Russian mathematician, Lobachevsky:

> *"The trouble with the Russian geometer's wheel, as well as with that of the other geniuses who employ this "infinite radius," is that the rubber tire is too far from the ball bearings for safety and convenience for handling."*

———¤¤¤⛝⛝⛝¤¤¤———

Most Determined Follower of Euclid' 5th Postulate

World Renowned Parallel Postulate by Matthew Ryan. (Henry Wilkins Co.) Wash. D.C., 1905.

Euclid had no follower as devoted as a Mr. Matthew Ryan, claiming anyone who fails to believe Euclid's fifth postulate is "a follower of Satan." Nothing is actually proven in this 29 page treatise, but believers of Lobachevsky's work are raked over the coals. A typical opinion that Mr. Ryan puts forth is:

> "...*anti-Euclidian geometry was founded on the false Satan assumption of the existence of non-Euclidian space.*"

The truly amazing thing about the proofs in this book is the fact the author makes no assumptions whatsoever. Schools also take a broadside from Mr. Ryan as he finishes the book with the following:

> "*The teaching of imaginary or non-Euclidian*

geometry in colleges and schools will breed an arrogant and imbecile race of students."

Mr. Ryan concludes with a generous thank you:

"I thank God, the Eternal Euclidian Geometer who commands the stars in their course has revealed to the author two demonstrations of the parallel postulate as a reward for forty-seven years of contemplation."

ΠΨΞθΦΧ

16

A CALENDAR FOR THE
NEXT 25,000,000 YEARS

We've all heard that time heals all wounds, that time is on our side, that time is money, that time is a-flying, and a million other adages about time. The only thing we don't know about time is what it is. Time is something we all experience every waking moment of our lives, but it's a conundrum encased in a puzzle inside a maze. So nebulous is the concept that we can't even experience it with our senses, even though we've been living and passing through it since the beginning of __ .

Although physicists have yet to satisfactorily answer the fundamental question, "What is time?," it has been measured from, uh, time immemorial by observing the motions of the heavenly bodies, which move with near constant regularity. It is due to the motions of the earth, moon, and sun that we measure the lengths of the day, month, and year, respectively. It is only the week, consisting of seven days, that was not created by the laws of nature, but by the laws of man. Although calendars can be based either on the sun or moon, the calendar adopted in Western culture is based on the sun.

In the 2nd century B.C., the Greek astronomer Hipparchus reckoned the length of time it took the earth to go from one summer solstice to the next summer solstice, or what we call one (solar) year, to be 365 days, 5 hours, 55 minutes and 12 second, which translates to 365,24667 days in a year. Of course, Hipparchus did not really have instruments precise enough to find that accurate an estimate, but the .24 after the decimal was spot on. Today, astronomers give a more precise number of days in a year as 365 days, 5 hours, 48 minutes, and 45.51 seconds, or 365.24219 days in a year.

The problem then arises, if a year is to be subdivided into 12 months, how many days should be allocated to each month? If we take the number of days in a year to be exactly 365, dropping the .24219 after the decimal, then (if you get out pencil and paper), you will discover the following nursery rhyme yields the correct number of days in a year.

Basic 365 Days per Year Calendar Rhyme

Thirty days hath September,
April, June and November,
All the rest have 31,
Except for February which has 28.

The problem with this calendar lies in the fact

that there are not 365 days in a year. If the above 365 days/year calendar is used, then after1 year it will be off by 6 hours, and after 4 years by 1 day, and after 4 × 180 = 720 years by 180 days, meaning that the calendar will predict snow in July and heat waves in January.

For this reason the Julian calendar was introduced.

Julian Calendar

The Julian calendar, introduced in 46 B.C. by the Roman Emperor Julius Caesar, is based on the more accurate approximation of 365.25 days in a year, written in fractional form as

$$365.25 = 365\ 1/4 \ \text{(Julian calendar)}$$

Hence, the Julian calendar adds an additional 1/4 th of a day each year, or equivalently one extra day every 4 years. Those years are called leap years, and the extra day is added to February, changing February from its customary 28 days to 29 in leap years. A year is a leap year if it is divisible by 4, like 2012, 2016, 2020, ... and so on.

More Calendars

Although the Julian calendar was a major improvement over earlier calendars, and the rule for leap years was easy for people to understand, it is slightly, about 11 minutes, too long, meaning over time, it too will get out of sync with the seasons.

For that reason, we use new and improved calendars. Of course, the gold standard of calendars would be to use the most accurate number of days in a year as measured by astronomers. This number is 365.24219 days/year, which in fractional form is

$$365.24219 = 365 \, \frac{24{,}219}{100{,}000} \quad \text{(accurate, not practical)}$$

This translates into adding 24,219 leap years every 100,000 years. But, to work out a scheme for doing this would be, well, ... Also, the calendar is more accurate than necessary and so we seek an adequate calendar that is easy to implement.

If we round 365.24219 to three places, or 365.242, we get the fractional equivalent

$$365.242 = 365 \frac{242}{1,000} \quad \text{(good calendar, not practical)}$$

which means adding 242 leap years every 1,000 years. No doubt we could implement a strategy for adding these leap years, but we have yet a better calendar, the Gregorian calendar.

The Gregorian calendar, decreed by Pope Gregory in 1582, is based on using 365.2425 days/year, which in fractional form is

$$365.2425 = 365 \frac{97}{400} \quad \text{(Gregorian calendar)}$$

This calendar has 97 leap years every 400 years and is easy to implement since it has 3 *less* leap years per 400 years than the Julian calendar, which has 100 leap years every 400 years. Hence, to implement the Gregorian calendar, we fine tune the Julian calendar by dropping three leap years every 400 years, which can be done by simply dropping the leap year which occurs on the first year of a century when the century is not divisible by 400. In other words, the years 1700, 1800, 1900, 2100, which are leap years in the Julian calendar, are not leap years in the Gregorian calendar, whereas 2000, 2400, 2800, ... which are leap years in the Julian calendar still leap years in the Gregorian calendar.

The Gregorian calendar is so accurate that it requires 75,000 years for the calendar to interchange summer and winter seasons.

New Age Calendar

If people ever become dissatisfied with the Gregorian calendar, one can always resort to my own favorite calendar, based on 365.2422 days in a year, a number very close to the astronomer's best estimate of 365.24219 days/year. In fractional form this number is

$$365.2422 = 365 \frac{872}{3600} \quad \text{(new age calendar)}$$

which means it adds 872 leap years every 3600 years, which is easy to implement since the Gregorian calendar adds 97 days every 400 years, or $9 \times 97 = 873$ days every 3600 years. Hence, we simply subtract one leap year from the Gregorian calendar every 3600 years, which we can do by dropping the Gregorian leap year that occurs on the first year of a century when the year is divisible by 3600. In other words, drop the Gregorian leap years 3600, 7200, ...

School children at that time will memorize the New Age calendar rhyme:

New Age Calendar

Thirty days hath September,

April, June and November,

All the rest have 31,

Except February which has 28.

Julian Amendment: Except when the year is divisible by 4, then it's a leap year.

Gregorian Amendment: If the first year of each century is not divisible by 400, the year no longer a leap year.

New Age Amendment: Drop the Gregorian leap years for years divisible by 3600, i.e. the years 3600, 7200, and so on.

There shouldn't be a great deal of criticism with this calendar since it will take 25,000,000 years for the calendar to exchange the seasons.

ΠψΔΦθΛ

17

THE INGENIOUS PIGEON HOLE CARD TRICK

———

The Great Zamboni, mentalist, card sharp, and part-time Rasputin impersonator has been scouring the land, appearing at strip malls and salesman award shows, where he performs his acclaimed Pigeonhole Card Trick. He is accompanied by the seductive Miss. Letticia, known for her alluring charms as well as her mathematical knowledge of the pigeonhole principle and expertise in permutations.

The author of this book, being an old acquaintance of Zamboni, has received permission to reveal the secret behind his famous turned-down card trick.

The illusion is impressive to a viewer and at the same time illustrates some interesting mathematical concepts. Horror of horrors, just when you thought you were going to learn some honest-to-goodness magic, it's all mathematics!.

The act begins when Miss. Letticia gives a deck of 52 playing cards to a volunteer in the audience who randomly selects five cards and then returns them to Miss. Letticia, Suppose the five cards chosen are shown in Figure 1.

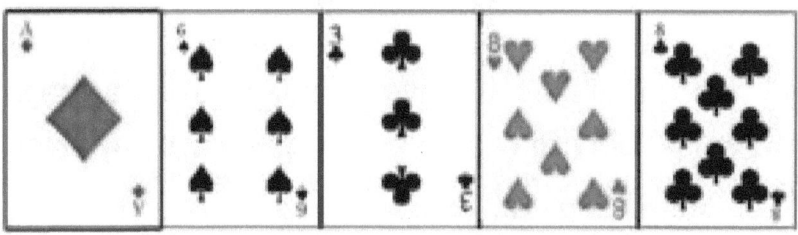

Figure 1: Miss Letticia receives these five cards.

Miss. Letticia then looks at the cards and places them with one card face down on a table for everyone to see. See Figure 2.

Figure 2: These cards encode the eight of clubs

The Great Zamboni, upon viewing the cards for the first time, goes through his usual mumbo-jumbo of hocus-pocus before reaching the proverbial aha" moment, whereupon he announces the turned-down card is an 8♣ — after which Miss. Letticia reveals the 8♣ to the audience.

—¤¤¤Ξ Ξ Ξ¤¤¤—

So, which of the following ruses did Zamboni and Miss. Letticia carry out to perform the trick? Take your pick. :

   1. Zamboni is a card-whisperer.

2. Miss. Letticia secretly passes information to Zamboni by nefarious eye motions.

3. Zamboni and Miss. Letticia each took Combinatorics 101 at a local community college.

You guessed it, the answer is c) elementary combinatorics.

—¤¤¤ΞΞ¤¤¤—

Two mathematical principles are used in this magic act — the pigeonhole principle and permutations. The pigeonhole principle states that if more than n pigeons try to squeeze in n pigeonholes, then at least two pigeons must share a bunk — or stated another way there are at least two people in New York City with the same number of hairs on their head.

The pigeonhole principle applied to our five cards states that at least two cards have the same suit, in our case two clubs, 3♣ and 8♣. Upon viewing these two cards, Miss. Letticia identifies the *smaller gap* between them as illustrated in the clock drawing in Figure 3. If you start at 8♣ and move clockwise to 3♣ the gap is eight, whereas if you start at 3♣ and move clockwise to 8♣, the gap is five. Selecting the smaller gap, we take 3♣ as the smaller club and 8♣ as the larger club. It is important to note that the smaller gap between any two cards will always be 1,,3,4,5 or 6.

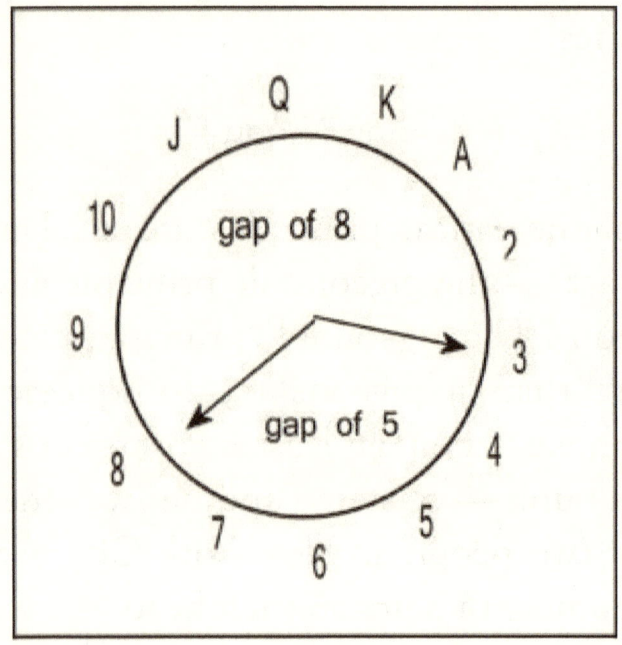

Figure 3: The 3 is the smaller club, the 8 the larger club

After identifying the smaller and larger clubs, Miss. Lettiicia places the larger 8♣ face-down (this is the mystery card) and the smaller 3♣ face-up next to it. The remaining three cards of 8♥, 6♠, A♦ are placed in left-to-right order as shown in Figure 2. The order seems unimportant to the observer but the order chosen by Miss. Letticia encodes the number five, which is the number of gaps between 3♣ and 8♣, thus allowing Zamboni to determine that the turned-down card is an 8♣.

So how does Miss Letticia's order (8♥, 6♠, A♦) encode the number five? The idea is quite simple provided Miss Letticia is can do some quick computations in her head. The idea is to order the entire 52-deck of cards from high to low by first ranking the suits by

$$\text{spaces} < \text{clubs} < \text{diamonds} < \text{hearts}$$

then within suits by

$$A < 2 < 3 < 4 < \ldots < 10 < J < Q < K$$

which orders the three cards according to 6♠ < A♦ < 8♥. Hence, Miss Letticia's placement of the cards in the order

$$(8♥, 6♠, A♦) \sim (\text{Largest, Smallest, Middle}).$$

We now assign a number 1, 2,..., 6 to each of the six permutations (or orders) of three cards as follows:

(Smallest, Middle, Largest) = 1

(Smallest, Largest, Middle) = 2

(Middle, Smallest, Largest) = 3

(Middle, Largest, Smallest) = 4

(Largest, Smallest, Middle) = 5

(Largest, Middle, Smallest) = 6

and so Miss Letticia's ordering of the cards (8♥, A♦ 6♣) encodes the number five, which tells Zamboni that the turned-down card is five larger than the 3♣ or 8♣.

—¤¤¤ΞΞ¤¤¤—

More than One Pair: If the five cards contained two hearts and two clubs, Miss. Letticia can work with either the hearts or clubs and continue as before.

However, suppose Miss. Letticia is presented

with five cards where three cards or more are of the same suit. Miss. Letticia can still determine a strategy with slightly more thought. Can the reader determine a strategy for these cases?

$$\Pi\Psi\Xi\theta\Phi X$$

18

THE SKELETON IN GOD'S CLOSET

"Yes?" the chief ombudsman said, looking up from his desk.

"May I help you?"

"Oh," the intruder paused at the door.

"Don't be afraid, I'm here to help."

"I've never come here before. I've never had anything to say, but now it can't wait," the man at the door said nervously.

"Fiddlesticks, just sit yourself down and tell me your problem." With that, the large pleasant-looking man behind the desk rose and nodded in the direction of a nearby chair. The man sat down and began to speak.

"I don't think you realize sir, but we have a problem."

"What kind of problem? By the way, what should I call you?"

"Willoby, sir, 67701."

"Well, Mr. Willoby, what is our problem?" the ombudsman, said as he pulled Willoby's dossier from a filing cabinet."

It's Hell."

"Hell?"

"Yes sir, Hell," Willoby paused, then stuttered

nervously, "uh ... it's ... it's ... getting better than ... uh .... Heaven."

"What?" the ombudsman said, "How can anything be better than here?"

"Well," Willoby said uneasily not looking up., "It all started with the mathematicians. Hell is full of them. There are topologists, algebraists, and even some, uh, applied mathematicians. Why, there are so many applied mathematicians and engineers down there they've installed air-conditioning. It's a pleasant 70 degrees. They say it's more comfortable there than here."

"*What?*" the ombudsman repeated, not believing his ears.

"It's true, it's true, and that's just the beginning, they,... "

"Wait Willoby, let me get this straight, you say that there are so many mathematicians in Hell they've got air-conditioning?"

"That's right," Willoby said, "I have this friend ... "

"All right, I believe you, you wouldn't be here if you were a liar."

Willoby smiled, he had led an exemplary life on earth and now in Heaven he was a model angel, but he felt it was his duty to inform God's chief ombudsman and troubleshooter about these latest goings-on at the other end of the Universe. There had been a lot of scuttlebutt around Heaven, but the rumors hadn't reached the top.

"That's right," Willoby continued, "but it's a lot more serious than the air-conditioning. Everyone has the right to make themselves comfortable. I understand it was awfully hot down there."

"I suppose," said the ombudsman.

"But, over the last few years other kinds of earthlings, besides these mathematicians, have gone to Hell after their life on earth."

"What kinds?" asked the ombudsman.

"Like I said, it all started with the mathematicians but later there were writers, painters, scientists, you name 'em, all kinds of people. In fact," Willoby cleared his throat, "even some, some, uh, theologians."

"Theologians?"

"It's true, it's true, and I can prove it."

"Ok, ok," the ombudsman interjected. "I believe you. But how could a theologian not get into Heaven?"

"That's just the point, sir," Willoby said. He stammered for a moment.

"Well?" the ombudsman waited anxiously.

"Well, you see sir, uh, they don't all want into Heaven anymore." Again, the ombudsman stared at Willoby in disbelief.

"What? Why would anyone go to Hell before Heaven? It's against the laws of the universe," the ombudsman said.

"They say Hell is more interesting than Heaven," Willoby continued. "The mathematicians hold weekly seminars and lectures down there. It's very exciting."

The ombudsman gasped.

"In fact," Willoby continued, starting to gain confidence, "during the past twenty-five years, architects, urban-planners, biologists, computer scientists you name 'em, they're all opting for Hell. They've designed complete cities using the latest technology. They have internet and can send e-mail messages from one end of Hell to the other. They've even installed a new computerized network and have their own website

http://www.come-on-down.hell

Computerized trains zip people from one end of Hell to the other. Botanists have even developed new strains of plants that grow in that awful soil. They even have Facebook! They e-mail kitten videos from one end of Hell to the other. They've turned it into a paradise, it's a,a,uh, a heaven."

"*Heaven!?*" the ombudsman shouted.

"Is that what they call it, a *Heaven*!?"

"Well, uh" Willoby tried to defend himself, then continued, "we're not getting the right kind of people anymore.

"You're saying we should change our admittance requirements?" the ombudsman asked.

"It wouldn't work, "Willoby said, "there are so many advertising executives down there that the instant an earthling dies, he's bombarded with a media blitz."

"*Nooooooo*!?" the ombudsman gasped.

"It's true, at the exact instant of death, a person gets an e-mail from Hell with a message like. *You're in good hands with Lucifer*, or *When you say Beelzebub you've said it all*, or even *With a name like Mephistopheles it has to be good.*"

The ombudsman was stunned. Something clearly had to be done to stop this brain drain to Hell.

After a few moments of silence, Willoby spoke again.

"There is still another aspect which I haven't mentioned, far more serious."

The ombudsman's face grew pale.

"More serious?" he gulped.

"Yes," Willoby said. "It concerns some angels here in Heaven."

"Go on."

Hmmmmmm. what's going on down there?

"Well," Willoby said, not looking directly into

the ombudsman's eyes, "some, uh, some want to be transferred."

"Transferred!?" the ombudsman shouted, jumping out of his chair. "Transferred? Is that what they call it!?"

"Some do," Willoby stuttered.

"That does it!" the ombudsman shouted. "This conversation is going straight to the Big Guy himself. We're going to nip this problem in the bud!" With that, he grabbed the red phone and dialed #1.

"Doris, is the Boss in? This is urgent," the ombudsman said, holding the receiver tight against his ear.

After a moment, he spoke.

"I see, oh really, that's strange. Well, let me know when he returns." With that, he swung around and looked intently at Willoby.

"What is it?" Willoby asked.

"Strange," the ombudsman said, "God's not in his office this afternoon and said he wouldn't be back until Monday." The ombudsman looked puzzled.

"What is it?" Willoby pressed.

"Well," the ombudsman said. "When he left the office, he said something strange."

"What?" Willoby asked.

"He just said

$$e^{i\pi} + 1 = 0$$

and walked out the door. He just said and walked out the door," he repeated. "*I wonder what he meant by that?*"

$$\Pi\psi\Delta\Phi\theta\Lambda$$

19

THE LEGEND OF FRANK NELSON COLE

Throughout history, hundreds of important mathematical discoveries (or *inventions* if you so care to describe them) have been made. A few of them include Euclid's Elements, Decartes' discovery of analytic geometry, the acceptance of the complex number i, the invention of abstract algebra in the late 19th century, Godel's Incompleteness Theorem, and on and on and on.

There are so many one could write a very large book detailing all of them. .

Now, I'm not one to brag, but I myself have performed an original mathematical operation, which I believe has never been carried out before in the annals of mathematics. As far as I know, I am the only person to have ever added the following two numbers:

$$503755543261043141546782394532002247$$
$$+253209346515604532251205302254099312$$
$$759964889776647673797987696786699559$$

Granted, many numbers have been summated throughout history, but as far as the author knows, no person has ever added these two *particular* numbers.

No doubt, there are skeptics among the reader who dispute my claim of adding two particular numbers, even if they have never been summed before, does not constitute anything special, let along a mathematical breakthrough. But *au contraire*, let me tell you about Frank Nelson Cole.

Back in the 17th century in a small French monastery, a monk by the name of Marin Mersenne was trying to find a formula to generate all the prime numbers. His goal failed but in the process he brought to attention numbers of the form $2^n - 1$, i.e. numbers that are one less than powers of two. Today numbers of this form are called Mersenne numbers, the first six Mersenne numbers listed below.

- $M(2) = 2^2 = 4 - 1 = 3 \leftarrow$ prime
- $M(3) = 2^3 = 8 - 1 = 7 \leftarrow$ prime
- $M(4) = 2^4 = 16 - 1 = 15 \leftarrow$ composite
- $M(5) = 2^5 = 32 - 1 = 31 \leftarrow$ prime
- $M(6) = 2^6 = 64 - 1 = 63 \leftarrow$ composite
- $M(7) = 2^7 = 128 - 1 = 127 \leftarrow$ prime

Mersenne proved that if the exponent of 2 is not a prime number (i.e. a composite) like 4, 6, 8, 9, ...

, then its Mersenne number is also composite, as illustrated by

- $M(4) = 2^4 - 1 = 15 = 3 \times 5$
- $M(6) = 2^6 - 1 = 63 = 3 \times 21$

However, when the exponent of 2 is a prime number, the Mersenne number may be either prime or a composite. For example, the smallest prime numbers 2, 3, 5, 7, 11, ... have prime Mersenne numbers

- $M(2) = 2$
- $M(3) = 7$
- $M(5) = 31$
- $M(7) = 137$

and are called **Mersenne primes**. However, the Mersenne number of the prime number 11 is not prime, as seen by

$$M(11) = 2^{11} - 1 = 2048 - 1 = 2047 = 23 \times 89$$

Now, you are probably thinking after looking at

the above numbers that most Mersenne numbers are prime, but if you thought that, you
would be very, very wrong. So far out of the tril-
lions of numbers searched by computers, only
49 Mersenne primes have been discovered, the
largest one being

$$M(74{,}207{,}281) = 2^{74{,}207{,}281} - 1$$

which was found in January, 2016. We have
decided not to print this number in the book
since it contains 22,338,618 digits and using the
12-point Times Roman font of this book, it
would require 35 miles of digits, or 12,000 pages
of text.

——¤¤¤ΞΞΞ¤¤¤——

So what does all this have to do with Frank Nelson Cole? In 1903 it had been proven that the
Mersenne number

$$M(67) = 2^{67} - 1$$

a number 21 digits long, was composite, but no
one had actually verified the proof by factoring

the number into its component parts, no small feat in 1903 with pencil and paper.

However, on October 31, 1903 Cole took the stage at a meeting of the *American Mathematical Society* where he approached the chalkboard and proceeded to compute

$$M(67) = 2^{67} - 1 = 147,573,952,589,676,412,927$$

He then moved to another board and wrote

$$193,707,721 \times 761,838,257,287$$

and methodically worked out the endless multiplication by hand, eventually arriving at the correct value for $M(67)$. Upon finishing his calculations, not uttering a single word during the hour-long presentation, he returned to his seat to a standing ovation. When someone asked him how long it took him to find the two factors, he replied "three years of Sundays."

In the history of mathematics, Frank Nelson Cole's verification that $M(67)$ is not prime, is the only mathematical talk ever given without uttering a single word.

JERRY FARLOW

ΠΨΞθΦΧ

20

IS IT A PUZZLE OR IS IT MATHEMATICS?

———

There was a man in our town who had the annoying habit of checking out the puzzle books in the library and writing down all the solutions in the front flap of the book. Then one day someone found him with an axe in his head and to this day it has been a real puzzle who did it.

The Oxford University dictionary defines a puzzle as a game, toy, or problem designed to test one's intelligence and ability to think outside the box. The origin of the word "puzzle" is unknown but goes back at least as far as the Old French word "pulse" which means "confound" or "bewilder."

Solving puzzles provide an enjoyable experience while at the same time stimulating the brain. They can aid in the development of a child's intellect as well as keeping active the mind of the elderly. Puzzles come in an endless variety, ranging from jigsaw puzzles, word puzzles, number puzzles, logical puzzles, visual puzzles, and dozens more.

Solving puzzles often forces one to think outside the box, but sometimes they require thinking so far outside they might be considered trick questions. For example, can you think outside the box far enough so you can make the following incorrect equation correct by drawing a single line somewhere in the equation?

$$5 + 5 + 5 + 5 = 5555$$

Give up? Why it's easy, simply draw line depicted in the following true equation.:

$$5 + 54 \cdot 5 + 5 = 5555$$

So is this a puzzle or a trick question? Certainly, those who solved it thought outside-the-box.

—¤¤¤ΞΞ¤¤¤—

One of my favorite puzzlists, and there have been several over the past few centuries like the American Sam Loyd, British Henry Dudeney and American Martin Gardiner, is the captivating logician Raymond Smullyan, who poses logical puzzles that will fry your brain. One of my favorite Smullyan conundrums is his Knights and Knave's Problem, which states:

Island of Knights and Knaves

An island consists of only knights and knaves, where the knights always tell the truth and the knaves always lie. Smullyan visits the island and meets three people, which he calls A, B, and C.

He is told that at least one of the three is a knight and at least one of the three is a knave. He is also told that one of the three has a prize and if Smullyan can determine which has the prize, he can have it. The three persons A, B, and C make the following statements to Smullyan:

- A: Person B doesn't have the prize.
- B: I don't have the prize.
- C: I have the prize.

The question is, who are the knights and knaves and who has the prize?

We leave it to the reader to ponder.

Answer: All problems start with assumptions. What makes this problem so nasty is you don't whether an assumption is true or false since each person may be telling the truth or lying!

However, a good strategy for problems like this is to systematically assume each of the three possible solutions (A,B, or C has the prize) and see what shakes out from each assumption.

Assume C has the prize: If person C has the prize, then all three statements are true, which means all three persons are knights, which is contrary to what is given. Hence C does not have the prize.

Assume B has the prize: If B has the prize, then all three statements are false, which means all there persons are knaves, which again is contrary to what is given. Hence, B does not have the prize.

Since we are given that one of the three has the prize, we conclude that A must have the prize and further that A and B are knights and C is a knave.

—¤¤¤ΞΞ¤¤¤—

Puzzle Solving or Mathematics?

Smullyan's Island Puzzle is interesting but is it only a puzzle or is it mathematics as well? On one level, of course it is mathematics since it involves assumptions and requires logical thinking similar to that used in solving mathematical problems. However, the solution of what we call "puzzles" is generally its only goal, whereas for a "mathematics" problem, although often logically similar to a "puzzle," is a problem embedded in a large body of mathematical knowledge. Once a mathematics problem is solved, that is more often than not the starting point for new problems and ideas, leading ultimately to the huge body of knowledge we call mathematics, linked by mathematical "problems" called theorems, and whose ultimate goal is to shed light on the world around us.

So let's hear it for puzzles.

$$\Xi \Omega \Theta Z \Psi \Pi$$

21

HOW I LEARNED TO STOP WORRYING AND LOVE THE PROOF

I'd like to tell you a few things about mathematical proofs you might have overlooked back in your school days. Mathematical proofs come in all shapes and sizes. There are proofs that mathematicians call "beautiful proofs", where each step of the verification elicits a response like interesting, clever, or wow, I never saw *that* coming, and so on. On the other hand, there are yeoman type proofs that get the job done, but that's it, proofs which elicit responses like ok, ok I get it, and so on.

There are also mathematical proofs, like the proof of Fermat's Last Theorem, which states that for all positive integers the equation

$$a^n + b^n = c^n$$

never holds for integer exponents $n = 3, 4, 5, \dots$. In this case the conclusion is not really important, but the steps used by English mathematician Andrew Wiles have led to new insights in mathematics. It is not surprising that if it takes 358 years to solve a problem, the methodology required to solve it breaks new ground.

On the flip side, there are proofs whose conclusion is important, like the product of two negative numbers is positive, but the steps required to prove it result are not overly inspiring.

Some proofs contain thousands of steps,

whereas some proofs only a few. In 1769 the Swiss mathematician Leonard Euler made the conjecture that it is *impossible* to find five positive integers a, b, c, d and e that satisfy the relation

$$a^5 + b^5 + c^5 + d^5 = e^5$$

This conjecture was not known to be true or false for 197 years until 1966 when mathematicians Lander and Parken, with the aid of their good friend, the CDC 6600, proved Euler's claim to be false with a one-equation counterexample:

$$27^5 + 84^5 + 110^5 + 133^5 = 144^5$$

This is clearly one of the shortest proofs in mathematics history, although you might want to tell that to the CDC 6600.

———¤¤¤ΞΞΞ¤¤¤———

A clever one-line proof is one that verifies 22/7 > π and is embodied in the single equation

$$0 < \int_0^1 \frac{x^4(1-x)^4}{1+x^2}\, dx = 22/7 - \pi$$

The integral is clearly positive and hence so is its value $22/7 - \pi$. The value of the integral can easily be found by expanding the numerator, dividing by $1 + x^2$, and then integrating each term — an exercise every student of freshman calculus should be able to carry out.

———¤¤¤ΞΞΞ¤¤¤———

There are proofs based on pictures that embody relevant ideas. One such visual proof verifies the conjecture that the sum of the first n odd numbers is n^2. In other words

$$1 + 3 + 5 + \cdots + (2n - 1) = n^2$$

Some examples are as follows:

$$1 = 1^2$$
$$1 + 3 = 2^2$$
$$1 + 3 + 5 = 3^2$$
$$1 + 3 + 5 + 7 = 4^2$$

This result can be proven rigorously by the process of mathematical induction, although the following drawing illustrates the result nicely.

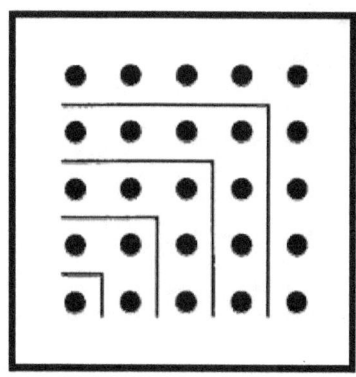

Can you "see" the identity?

However, one must be cautious when using visual proofs. The following optical illusion illustrates this point.

Do you see what's wrong here? Hint: You may

think you are looking at two triangles, but if you did you would be wrong. The "hypotenuse" of both the bottom and top figures are not straight lines. The top phony hypotenuse is "bowed inward" whereas the phony hypotenuse of the bottom figure is bowed outward, allowing the one extra square to squeeze its way into the bottom figure. In order for the hypotenuse of the largesy triangle to be a straight line, the ratios of the sides of the two smaller triangles must be the same, but one ratio is 2/5 whereas the other is 3/7, clearly not the same.

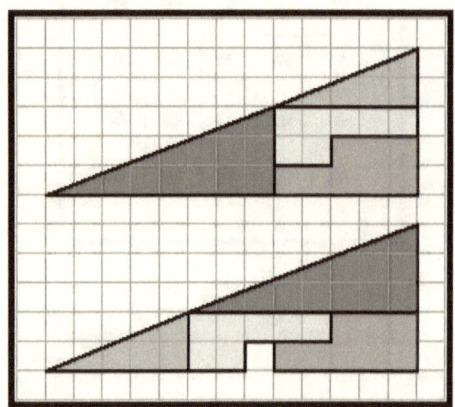

Be careful when using visual proofs

This paradox is called the Curry Paradox (or

missing box paradox), named after the magician, Paul Curry.

ΠΨΞΘΦΧ

22

THE AMAZING NUMBER 9

─────────

When I was but a wee tot, I was given an old hand-me-down 45 RPM record player along with a stack of equally-old 45 RPM records. My favorite song amongst the lot was *The Teddy Bear's Picnic*, a song that I played so frequently and so loud that it was my dear *mother* who threatened to run away from home. But thankfully for everyone involved the old vinyl spinner couldn't stand up to my melodic obsession with bears and ended up in the 45 RPM spin-after.

Upon hearing the "bear" song, as my mother referred to it, it was her cue to remind me that if I spent half the time learning my multiplication tables as listening to that senseless song about wild-animal picnics in the woods, I would become a mathematical savant. She would then try to stimulate my numeric lethargy by reciting some standard childhood learning poesy, like

6 times 8

Fell off the plate

That's what makes it 48.

Although even as an eight-year old, I scoffed at such infantile memes of learning, I have to admit that if you ask me today for the value of 6 × 8, the aforementioned jingle is apt to show signs of revival somewhere in the recesses of my brain.

Although learning multiplication tables is not a pleasurable experience for most grade-school students, there are many fascinating facts about basic arithmetic to stimulate the inquisitive mind. For example, did you know

- If you multiply any number 1,2,..., 9 by 9, the sum of the 10's and 1's digits of the product add up to 9 as in

$$6 \times 9 = 54 \rightarrow 5+4 = 9$$

$$3 \times 9 = 27 \rightarrow 2+7 = 9$$

- If you multiply any even number 2, 4, ... by 6, the last digit of the product is always an even number as in

$$8 \times 6 = 48$$

$$554 \times 6 = 3324$$

$$99448 \times 6 = 596688$$

and it's even more fascinating trying to understand *why* they're true!

I once knew a psychic who confided in me that the number 9 contained the all the mysteries of the universe and that it embodied spiritual symbolism that couldn't be deciphered by rational thought. (I though I'd tell him maybe it was because he lacked some of those quantities, but I though it best to bite my tongue.)

Mystery of the Circle One example exhibiting crop-circle goings on with the number 9 involves the innocent-looking circle, no doubt a shape you never suspected of harboring any wizardry qualities, other than the usual ones of causing mental angst to beginning geometry students.

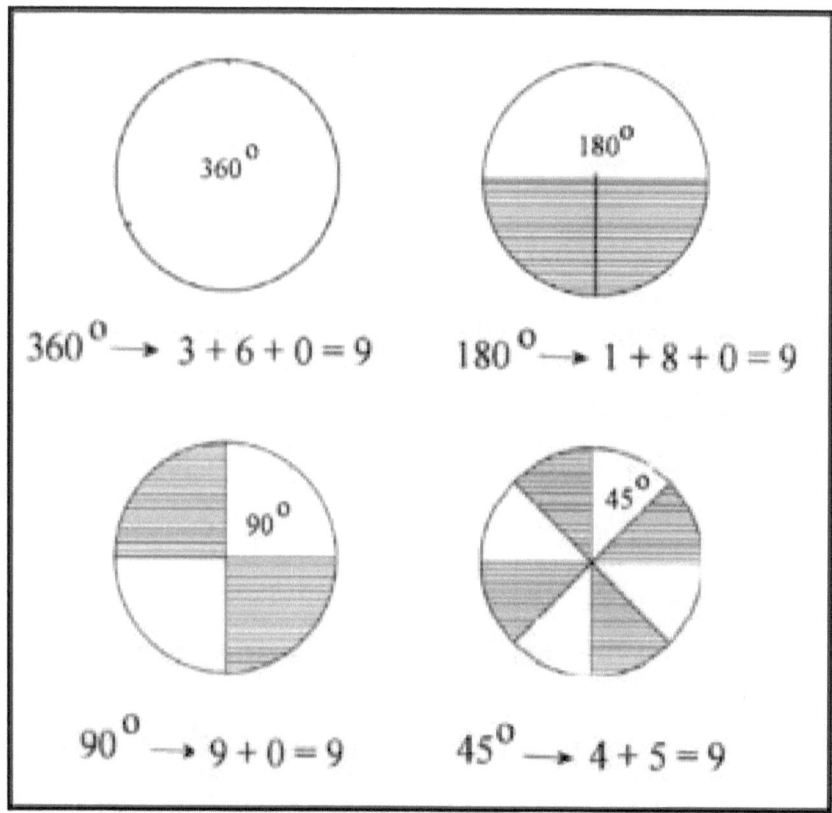

Hidden mysteries in a circle?

We all agree that it takes 360 degrees to go all the way around a circle and if you add the digits of 360 you get a 9, as in 3+6+0=9. Nothing out of the ordinary, but if you cut the circle in half, you get 180 degrees, whose digits also sum to 9. Coincidence? Possibly, but if divide the angle again you get 90 degrees and 9 + 0 = 9. Do it again! And again! Again! Look what happens!

$360/2 = 180 \rightarrow 1+8+0 = 9$

$180/2 = 90 \rightarrow 9 + 0 = 9$

$90/2 = 45 \rightarrow 4+5 = 9$

$45/2 = 22.5 \rightarrow 2+2+5 = 9$

$22.5/2 = 11.25 \rightarrow 1+1+2+5 = 9$

$11.25/2 = 5.625 \rightarrow 5+6+2+5 = 18 \rightarrow 1+8 = 9$

Even when the angle is less than one, its digits still sum to 9. There's something going on here !

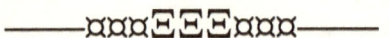

The Circle Magic Revealed

I hate to bring you the bad news but there are no hidden mysteries related to the number 9, just old-fashioned arithmetic. It has to do with the fact that

if a number is divisible by 9, then the sum of its digits is also divisible by 9.

To understand this strange property consider an arbitrary 3-digit number $d_2 d_1 d_0$ which can be written

$$d_2 d_1 d_0 = 100\, d_2 + 10\, d_1 + d_0$$

$$= (99 + 1)\, d_2 + (9 + 1)\, d_1 + d_0$$

$$= 99\, d_2 + 9\, d_1 + (d_2 + d_1 + d_0)$$

Assuming the number is divisible by 9, or equivalently is a multiple of 9, the above number can be written as

$$99\, d_2 + 9\, d_1 = (d_2 + d_1 + d_0) = 9N$$

where N is some positive integer. If we now solve for the sum of the digits of $d_2 d_1 d_0$ we get

$$d_2 + d_1 + d_0 = 9\, N - 99\, d_2 - 9\, d_1$$

$$= 9\,(N - 9\,d_2 - d_1)$$

$$= 9\,M$$

where

$$M = N - 9\,d_2 - d_1$$

clearly a positive integer. But this statement says the sum of the digits of $d_2 d_1 d_0$ is a multiple of 9 and hence divisible by 9.

The above result may not excite the average reader but it should. You see, (this is important) since the sum of the digits is divisible by 9, then so is the sum of the digits of the *sum of the digits*, and if we continue this process again and again we will arrive at a single digit number divisible by 9, and this digit must be 9. This digit is called the DIGIT SUM of the original number and it is 9 if the original number is divisible by 9. In other words *a number that is divisible by 9 has a DIGIT SUM of 9.*

The following example illustrates this process

summing the digits of a number divisible by 9 two times before arriving at its DIGIT SUM of 9.

$$77376627 \rightarrow 7+7+3+7+6+6+2+7 = 45 \rightarrow 4+5 = 9$$

To get a better grasp of the DIGIT SUM of a number, it turns out that every positive integer, be it small ones like 8 or 25 or a large one like 230569234, can be "boiled down" to a single digit, either 1,2,3,4,5,6,7,8, or 9. It is so easy to find the youngest child can find it. It is found by simply summing the digits of the number. If this sum is larger than a single digit, then one repeats the process again, and so on until arriving at a single digit. This single digit is called the DIGIT SUM of the given number. For example the DIGIT SUM of 34 is 3+4=7. The digit sum of 589 is 5 + 8 + 9 = 22, and since this number is more than a single digit, we repeat the process again getting the DIGIT SUM of 2 + 2 =4. The following numbers illustrate a few more DIGIT SUMS.

Number	DIGIT SUM
7	7
26	2+6=8
589	5+8+9=22$\rightarrow$2+2=4
9999	9+9+9+9=36$\rightarrow$3+6=9

Note that of the above numbers only 9999 has a DIGIT SUM of 9, meaning it is the only number among the group that is divisible by 9. The DIGIT SUMS of the other numbers are the remainders when the given number is divided by 9. For example the number 7 has remainder 7 when divided by 9, 26 has remainder 8, and 589 has a remainder of 4 (which means 9 divides 585). (Check them yourself).

The Circle Mystery

Finally! Now that we know that any number divisible by 9 has a DIGIT SUM of 9, we can explain the mystery of the circle. Starting with 360 degrees in a circle, we repeatedly divide the circle by 2 getting the following angles:

$360/2 = 180 \rightarrow 1 + 8 + 0 = 9$

$180/2 = 90 \rightarrow 9 + 0 = 9$

$90/2 = 45 \rightarrow 4 + 5 = 9$

$45/2 = 22.5 \rightarrow 2 + 2 + 5 = 9$

$22.5/2 = 11.25 \rightarrow 1 + 1 + 2 + 5 = 9$

$11.25/2 = 5.625 \rightarrow 5 + 6 + 2 + 5 = 18 \rightarrow 1 + 8 = 9$

$5.625.2 = 2.8125 \rightarrow 2 + 8 + 1 + 2 + 5 = 18 \rightarrow 1 + 8 = 9$

We know that any number, such as 360, that is divisible by 9 has a DIGIT SUM of 9, but many of the above fractional parts are not divisible by 9. The reason they have DIGIT SUMS of 9 however has to do with the way we can express them as

$360/2 = 5 \times 360/10$

$360/4 = 25 \times 360/100$

$360/8 = 125 \times 360/1000$

$360/16 = 600 \times 360/10000$

Since 360 is divisible by 9, then any multiple of 360, like 5×360, 25×360, 125×360 is also divisible by 9, and hence has a DIGITAL SUM of 9. But what about the 10, 100, ... and so on in the denominator? These divisors only shift the decimal point of the number and do not change the DIGITAL SUM of the number. Hence all the numbers have DIGITAL SUMS of 9. No magic here, just simple arithmetic.

ΣΛΘΔΓΞ

23

AN OLD COLLEGE TEXTBOOK YOU MIGHT RECALL

———

Sometimes I wake up at night drenched in sweat. Far off, I hear faint cries of tormented students. I am well aware of anguish endured by hundreds upon hundreds of these tortured souls. Then, it strikes me, maybe they know the whereabouts of where I lie, and now their cries grow louder and louder. They are coming for me by the hundreds, thousands, ... My heart is pounding like a drum. They're getting CLOSER, CLOSER! Then I think of the massive royalties that keep rolling in, and I turn over and sleep like a baby.

One thing I've always eschewed, well, maybe not eschewed, but avoided like the plague, is the fact that I might be the person that wrote those overpriced textbooks that sucked your bank account dry back in your college days. You know, all that hard-earned cash you expected to go for booze and wild parties, went instead to buy my seminal text, Adventures with the Quadratic or my all-time favorite, A Passion for the Polynomial..

It gives me no pleasure knowing that at any hour of the day there are thousands of poor college freshmen using my name in vain.

"*What in the hell is this guy talking about?*" is probably running through the minds of countless numbers at this very moment. Or no doubt "BORING," is also a common theme. And, of course, the always popular, "*Where the hell are the answers to the problems?*" That, of course, is a dumb question. They should know by now if they want the answers to the problems, it'll cost them an extra $59.95. Did they actually think the answers came with the textbook? *Ha, ha, ha ,... .*

Why did I put my address in that book?

I have been called the Stephen King of Algebra. Students experience more terror on the first page of my *Adventures with the Quadratic* than in a Stephen King novel. For that reason, I would like to share with you a few secrets I've developed over the years.

Rule 1: The Used Book Paradox

If there's one thing of which I take umbrage, it's the humongous line of students that form at the college bookstore at the end of the semester. They, of course, are carrying out the unspeakable act of *reselling* my books. Although the extra money may come in handy, I contend that keeping the textbook has several benefits. First, the student is able to reread the book at a later date, developing new insights. Secondly, the student adds another book to his mathematical library. And third, although not important but I mention it anyway, it knocks a hole in my royalty check big enough to drive a Mack truck.

To help the beginning student avoid the cheap lure of the resale counter, I have devised several effective strategies. One particularly effective method is my design of an ornate book plate, embellished on the cover of the book where the students can sign their name, campus address, and other relevant information. This information will be noticed by anyone finding the book if accidentally misplaced. It will also be noticed by the bookstore manager, classifying it as worthless junk.

Rule 2: Textbook Supplements

To assist schools who adopt my books, I provide instructors with a long line of pedagogical "extras," which includes *sample tests, computer software, instructor manuals, solution manuals, teaching guidelines, sample syllabi,* ... after which the school will have become so enmeshed in a maze of supplements that to change the book would require the skills of a licensed accountant.

Rule 3: Proper Font Size

A major expense of producing a new book is the cost of paper. For that reason, I avoid printing equations, symbols and tables in excessively large font. There is little reason to use the blatantly large type for the equation

$$y = \frac{3x^4}{z} + \tan^{-1}x + \frac{1}{(e^z + e^{-z})} + \sin y + 2$$

when the perfectly smaller size

$$y = \frac{3x^4}{z} + \tan^{-1}x + \frac{1}{(e^z+e^{-z})} + \sin y + 2$$

will suffice.

$$\Pi\psi\Delta\Phi\theta\Lambda$$